# An Introduction to Bond Graph Modeling with Applications

# An Introduction to Bond Graph Modeling with Applications

J. A. Tenreiro Machado

Dept. of Electrical Engineering, Institute of Engineering,
Polytechnic Institute of Porto

Vitor M. R. Cunha

Dept. of Electrical Engineering, Institute of Engineering,
Polytechnic Institute of Porto

CRC Press is an imprint of the
Taylor & Francis Group, an **informa** business

A CHAPMAN & HALL BOOK

First edition published 2021
by CRC Press
6000 Broken Sound Parkway NW, Suite 300, Boca Raton, FL 33487-2742

and by CRC Press
2 Park Square, Milton Park, Abingdon, Oxon, OX14 4RN

© 2021 J. A. Tenreiro Machado and Vitor M. R. Cunha

CRC Press is an imprint of Taylor & Francis Group, LLC

Reasonable efforts have been made to publish reliable data and information, but the author and publisher cannot assume responsibility for the validity of all materials or the consequences of their use. The authors and publishers have attempted to trace the copyright holders of all material reproduced in this publication and apologize to copyright holders if permission to publish in this form has not been obtained. If any copyright material has not been acknowledged please write and let us know so we may rectify in any future reprint.

Except as permitted under U.S. Copyright Law, no part of this book may be reprinted, reproduced, transmitted, or utilized in any form by any electronic, mechanical, or other means, now known or hereafter invented, including photocopying, microfilming, and recording, or in any information storage or retrieval system, without written permission from the publishers.

For permission to photocopy or use material electronically from this work, access www.copyright.com or contact the Copyright Clearance Center, Inc. (CCC), 222 Rosewood Drive, Danvers, MA 01923, 978-750-8400. For works that are not available on CCC please contact mpkbookspermissions@tandf.co.uk

*Trademark notice*: Product or corporate names may be trademarks or registered trademarks and are used only for identification and explanation without intent to infringe.

---

**Library of Congress Cataloging-in-Publication Data**

---

Names: Machado, J. A. Tenreiro, author. | Cunha, Vitor M. R., author.
Title: An introduction to bond graph modeling with applications / J.A. Tenreiro Machado, Dept. of Electrical Engineering, Institute of Engineering, Polytechnic Institute of Porto, Vitor M.R. Cunha, Dept. of Electrical Engineering, Institute of Engineering, Polytechnic Institute of Porto.
Description: First edition. | Boca Raton : Chapman & Hall/CRC Press, 2021. | Includes bibliographical references and index.
Identifiers: LCCN 2021000968 (print) | LCCN 2021000969 (ebook) | ISBN 9780367523411 (hardback) | ISBN 9781003057741 (ebook)
Subjects: LCSH: Bond graphs. | Bond graphs--Problems, exercises, etc. | Dynamics.
Classification: LCC TA338.B6 M33 2021 (print) | LCC TA338.B6 (ebook) | DDC 620.001/5115--dc23
LC record available at https://lccn.loc.gov/2021000968
LC ebook record available at https://lccn.loc.gov/2021000969

---

ISBN: 978-0-367-52341-1 (hbk)
ISBN: 978-0-367-52405-0 (pbk)
ISBN: 978-1-003-05774-1 (ebk)

Typeset in CMR10
by KnowledgeWorks Global Ltd.

# Contents

| | | |
|---|---|---|
| Preface | | ix |
| CHAPTER 1 ▪ Mathematical Models of Dynamical Systems | | 1 |
| 1.1 | INTRODUCTION | 1 |
| 1.2 | MODELING OF ELECTRICAL SYSTEMS | 2 |
| 1.3 | MODELING OF MECHANICAL SYSTEMS | 7 |
| | 1.3.1 Translation Motion | 8 |
| | 1.3.2 Rotational Motion | 11 |
| 1.4 | TRANSMISSION SYSTEMS | 14 |
| | 1.4.1 Conversion between Rotational Motions | 15 |
| | 1.4.2 Conversion from Rotational to Translational Motion | 16 |
| | 1.4.3 Translational to Translational Motion Transformation | 18 |
| | 1.4.4 Permanent Magnet DC Motors | 19 |
| 1.5 | MODELING OF HYDRAULIC SYSTEMS | 20 |
| 1.6 | MODELING OF THERMAL SYSTEMS | 25 |
| CHAPTER 2 ▪ Bond Graph Modeling | | 27 |
| 2.1 | INTRODUCTION | 27 |
| 2.2 | BASIC CONCEPTS | 27 |
| 2.3 | BOND GRAPH ELEMENTS | 30 |
| | 2.3.1 One-Port Elements: Energy Sources | 31 |
| | 2.3.2 One-Port Elements: Energy Stores | 32 |
| | 2.3.3 One-Port Elements: Dissipators | 33 |
| | 2.3.4 Two-Port Elements | 34 |

|        |       | 2.3.5   | Junctions                                              | 35 |
|--------|-------|---------|--------------------------------------------------------|----|
|        |       | 2.3.6   | Modulated Elements                                     | 36 |
|        | 2.4   | EXAMPLES OF APPLICATION OF BOND GRAPHS                 || 40 |
|        |       | 2.4.1   | Example 1: Electrical Circuit                          | 40 |
|        |       | 2.4.2   | Example 2: Translational Mechanical System             | 41 |
|        |       | 2.4.3   | Example 3: Hydraulic System                            | 42 |
|        |       | 2.4.4   | Example 4: Thermal System                              | 43 |
|        | 2.5   | CAUSALITY                                              || 44 |
|        |       | 2.5.1   | Causality: Energy Sources                              | 46 |
|        |       | 2.5.2   | Causality: Energy Stores (Integral Causality)          | 47 |
|        |       | 2.5.3   | Causality: Dissipators                                 | 48 |
|        |       | 2.5.4   | Causality: High-Order Elements                         | 48 |
|        |       | 2.5.5   | Causality: Junctions                                   | 49 |
|        |       | 2.5.6   | Causality: Two-ports                                   | 49 |
|        | 2.6   | ASSIGNING CAUSALITY TO A BOND GRAPH                    || 50 |
|        |       | 2.6.1   | Example of Causality Assignment: Electrical Circuits   | 52 |
|        |       | 2.6.2   | Example of Causality Assignment: Translational Mechanical Systems | 53 |
|        |       | 2.6.3   | Example of Causality Assignment: Hydraulic System      | 55 |
|        |       | 2.6.4   | Example of Causality Assignment: Thermal System        | 55 |

CHAPTER 3 ▪ Electrical Systems — 57

| 3.1 | INTRODUCTION       | 57 |
|-----|--------------------|----|
| 3.2 | SOLVED PROBLEMS    | 58 |
| 3.3 | PROPOSED EXERCISES | 66 |

CHAPTER 4 ▪ Mechanical Systems — 75

| 4.1 | INTRODUCTION       | 75 |
|-----|--------------------|----|
| 4.2 | SOLVED PROBLEMS    | 78 |
| 4.3 | PROPOSED EXERCISES | 86 |

| CHAPTER 5 ▪ Hydraulic Systems | 105 |
|---|---|
| 5.1 INTRODUCTION | 105 |
| 5.2 SOLVED PROBLEMS | 105 |
| 5.3 PROPOSED EXERCISES | 114 |

| CHAPTER 6 ▪ Thermal Systems | 127 |
|---|---|
| 6.1 INTRODUCTION | 127 |
| 6.2 SOLVED PROBLEMS | 127 |
| 6.3 PROPOSED EXERCISES | 134 |

| CHAPTER 7 ▪ Multi-domain Systems | 143 |
|---|---|
| 7.1 INTRODUCTION | 143 |
| 7.2 SOLVED PROBLEMS | 143 |
| 7.3 PROPOSED EXERCISES | 150 |

| CHAPTER 8 ▪ Bond Graph Modeling and Simulation Using 20-sim | 171 |
|---|---|
| 8.1 INTRODUCTION | 171 |
| 8.2 GUIDED EXERCISE | 172 |

| Solutions | 197 |
|---|---|
| Appendix A | 245 |
| Appendix B | 249 |
| References | 251 |
| Index | 255 |

# Preface

The idea for writing this book arises in the follow-up of our classes and various lectures on topics such as 'System Theory', 'Modeling and Control of Dynamical Systems' and 'Electromechanical Systems' held during several decades. The motivation for this topic started in the 80s, in the scope of research focused on the dynamical modeling of robotic manipulators driven by electrical and hydraulic actuators. Later, the interest in the bond graph modeling tool extended beyond technological aspects and robotics. The perspective of bond graphs provided a broader interpretation on the general aspects of mathematical modeling. By the end of the 90s we introduced the bond graph method to students attending classes of 'Electromechanical Systems'. At first we were not sure about the students receptivity to this mathematical tool, when presented in the middle of chapters such as hydraulics, robotics, mechanical transmissions and electrical motor control. To our surprise, the student receptivity was very good. During the succeeding years, we noted that the bibliography was somewhat scattered and was not friendly for students initiating their work with this tool. Therefore, we wrote several lecture notes and slides to highlight a variety of details. These notes and exercises were the leitmotif for working on a book covering the distinct topics written along the years and scattered along notes, exams and laboratory work guides, or simply discussed during the weekly classes. We are pleased that CRC Press is be publishing this book involving both the theory and exercises of bond graph modeling. We hope that students and other readers find this work useful.

J. A. Tenreiro Machado and Vitor M. R. Cunha
Institute of Engineering, Polytechnic of Porto, December 2020

CHAPTER 1

# Mathematical Models of Dynamical Systems

## 1.1 INTRODUCTION

In the study of dynamical systems, both for understanding, describing, predicting and controlling its behavior, we need a model. In general, a mathematical model consists of a set of differential equations that try to represent the dynamics of the system reasonably well. However, a mathematical model is not unique. A given system may be represented in different ways and, therefore, may have several mathematical models, depending on the adopted approach and the final purposes. Hereafter, we consider systems described by a set of ordinary differential equations that may be obtained by following the governing physical laws. The analysis and design of control systems for many applications use linear models, or eventually linearized versions of nonlinear models, not only because they are well established, but also because the treatment of nonlinear systems is considerably more complex.

In this textbook we consider that the principle of causality apply, meaning that the output of the system (i.e., the output at the time instant $t = 0$) depends on the past input (i.e., $t \leq 0$) but does not depend on the future (i.e., $t > 0$). Moreover, it is assumed that readers have the fundamental knowledge of Laplace transform, transfer functions and block diagrams.

The Laplace transform is given by:

$$\mathscr{L}\{f(t)\} = \int_{0^-}^{+\infty} f(t)e^{-st}\mathrm{d}t, \qquad (1.1)$$

where $f$ is a signal evolving in time $t$, $\mathscr{L}$ denotes the Laplace operator and $s = \sigma + j\omega$ stands for the Laplace variable with $j = \sqrt{-1}$.

The transfer function between the input and output signals, $x(t)$ and $y(t)$, corresponds to the ratio $\frac{Y(s)}{X(s)}$, where $X(s) = \mathscr{L}\{x(t)\}$ and $Y(s) = \mathscr{L}\{y(t)\}$ and zero initial conditions are assumed.

Hereafter, the notations $\dot{x}$ and $\ddot{x}$, or $\frac{dx}{dt}$ and $\frac{d^2x}{dt^2}$, will be considered interchangeably. Identically, if nothing is said, a variable $x$ denotes its time variation $x(t)$.

## 1.2 MODELING OF ELECTRICAL SYSTEMS

The modeling of electrical systems describes the relationship between two variables, namely the voltage $v$ and the current $i$, so that its scalar product corresponds to power $P$ (i.e., $v \cdot i = P$).

The Faraday's law establishes the relation between the voltage $v$ and the magnetic flux $\lambda$:

$$v = \frac{d\lambda}{dt}. \tag{1.2}$$

The current $i$ is related to the electrical charge $q$ by the equation:

$$i = \frac{dq}{dt}. \tag{1.3}$$

Alternatively, these equations can be written in the integral form so that:

$$\lambda = \int_0^t v(\tau) d\tau, \tag{1.4}$$

and

$$q = \int_0^t i(\tau) d\tau. \tag{1.5}$$

Electrical circuits are governed by the so-called Kirchhoff's laws. The current law (often also called first Kirchhoff law) states that the algebraic sum of all currents flowing towards, or away, from a node is zero:

$$\sum_{k=1}^{n} i_k(t) = 0, \tag{1.6}$$

where $n$ denotes the total number of branches connected to the node, and $i_k$ stands for the current flowing through the $k$-th branch.

The Kirchhoff voltage law (also called second Kirchhoff law) states that the algebraic sum of the voltages around a loop of the electrical circuit is zero:

$$\sum_{k=1}^{n} v_k(t) = 0, \quad (1.7)$$

where $n$ denotes the total number of voltages in the branches (or differences of potential between the terminals), $v_k$, around the loop.

The standard graphical representation of voltage drops and currents is designed by means of curved and straight arrows, respectively.

Electrical systems have two possible types of sources of energy, namely of voltage and current as represented in the diagram of Figure 1.1.

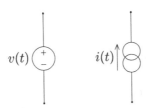

Figure 1.1  Electrical sources of energy: voltage $v$ and current $i$.

We have three basic constitutive equations of the elements that relate the voltage drop $v$, or voltage across its two terminals, and the current $i$ through the elements. The basic elements are called the resistor, inductance and capacitor that are described by:

$$v(t) = Ri(t), \quad (1.8)$$

$$v(t) = L\frac{di(t)}{dt}, \quad (1.9)$$

$$i(t) = C\frac{dv(t)}{dt}. \quad (1.10)$$

The symbols $R$, $L$ and $C$ stand for electrical resistance, inductance and capacitance. The three basic electrical elements are represented as depicted in Figure 1.2.

Expressions (1.9) and (1.10) relate $v(t)$ and $i(t)$ in the differential form, but we can write them in the integral form:

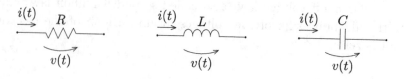

Figure 1.2 Basic electrical elements: resistance $R$, inductance $L$ and capacitance $C$.

$$i(t) = \frac{1}{L}\lambda + i(0) = \frac{1}{L}\int_0^t v(\tau)\mathrm{d}\tau + i(0), \qquad (1.11)$$

$$v(t) = \frac{1}{C}q + v(0) = \frac{1}{C}\int_0^t i(\tau)\mathrm{d}\tau + v(0), \qquad (1.12)$$

where $i(0)$ and $v(0)$ represent the current and voltage at $t = 0$, or simply, the initial conditions.

The voltage and the current can be seen as an *across* and a *through* variable, respectively. The term *across* variable is often used for voltage drop or difference of potential between the two terminals of the element. The term *through* variable is used for the current flowing within the element. The voltage $v$ and the current $i$ are also called *effort* and *flow*. The assignment of across or effort and through or flow variables is arbitrary, and we can consider the alternative without significant influence in the final mathematical model.

The resistance dissipates energy (the so-called Joule effect $W = Ri^2$), while the capacitance and the inductance store energy in the electrostatic and magnetic forms ($W = \frac{1}{2}Cv^2$ and $W = \frac{1}{2}Li^2$), respectively. The units of the variables and elements are given in Table A.2.

We can obtain the differential equations for modeling any electrical circuit in the time domain, by means of the Kirchhoff's laws (1.6)–(1.7) and the expressions (1.8)–(1.12).

The expressions (1.9)–(1.12) can be written in the Laplace domain. For zero initial conditions we have:

$$V(s) = RI(s), \qquad (1.13)$$

$$V(s) = sLI(s) \Leftrightarrow I(s) = \frac{1}{sL}V(s), \qquad (1.14)$$

$$I(s) = sCV(s) \Leftrightarrow V(s) = \frac{1}{sC}I(s). \qquad (1.15)$$

Figure 1.3 shows an example of a electrical circuit with input and output voltages $v_i$ and $v_o$, respectively, with one resistance $R$, one inductance $L$ and one capacitance $C$ in series.

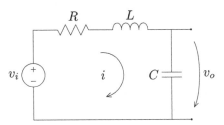

**Figure 1.3** Electrical circuit: resistance $R$, inductance $L$ and capacitance $C$ in series.

Applying the Kirchhoff voltage law we can write the mathematical model:

$$v_i = Ri + L\frac{di}{dt} + \frac{1}{C}\int_0^t i(\tau)d\tau + v(0), \qquad (1.16)$$

$$v_o = \frac{1}{C}\int_0^t i(\tau)d\tau + v(0), \qquad (1.17)$$

where $i$ is the current flowing through the three electrical elements.

By assuming zero initial conditions, we can obtain the transfer function, yielding:

$$\frac{V_o(s)}{V_i(s)} = \frac{1}{LCs^2 + RCs + 1}. \qquad (1.18)$$

Figure 1.4 shows an example of a electrical circuit with the input current $i_i$ and one resistance $R$, inductance $L$ and capacitance $C$ in parallel.

Applying the Kirchhoff current law we can write the mathematical model:

$$i_i = \frac{1}{R}v + \frac{1}{L}\int_0^t v(\tau)d\tau + C\frac{dv}{dt} + i(0). \qquad (1.19)$$

The three basic elements $R$, $L$ and $C$ are often called one-port elements since they establish a relationship between the voltage across (or

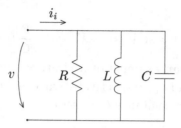

**Figure 1.4** Electrical circuit: resistance $R$, inductance $L$ and capacitance $C$ in parallel.

voltage drop between) their terminals and the current flowing through them. However, often, electrical systems include two-port devices that include a primary and a secondary circuit. Two-port components are power conserving and can be of two types, namely the transformer and the gyrator.

These devices adapt the voltage and current between the primary and secondary circuits. In electrical systems the transformer has a simple technological implementation by means of a magnetic circuit, Figure 1.5. On the other hand, the implementation of gyrators requires the adoption of operational amplifiers.

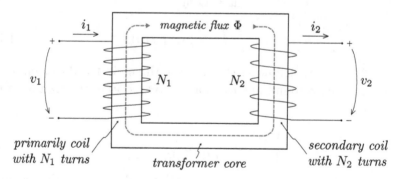

**Figure 1.5** Simplified technological implementation of an electrical transformer.

Let us consider the transformer represented on Figure 1.6 with voltages and currents $v_1$ and $i_1$ in the primary and $v_2$ and $i_2$ in the secondary circuit. Let us also consider that all magnetic flux $\Phi$ follows the magnetic core and is captured by the coils with $N_1$ and $N_2$ turns in the primary and in the secondary circuit. From the Faraday's law we obtain $\frac{v_1}{v_2} = \frac{N_1 \frac{d\Phi}{dt}}{N_2 \frac{d\Phi}{dt}}$ and assuming an ideal system we have that the power $P$ at input and output are identical, that is $v_1 \cdot i_1 = v_2 \cdot i_2$. Therefore, we can write the equations:

$$\frac{v_1}{v_2} = \frac{i_2}{i_1} = n, \qquad (1.20)$$

where $n = \frac{N_1}{N_2}$ is called the transformation ratio.

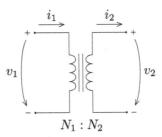

Figure 1.6  Diagram of an ideal electrical transformer.

The gyrator is represented schematically as depicted in Figure 1.7. For this device we have the equations:

$$\frac{v_1}{i_2} = \frac{v_2}{i_1} = r, \qquad (1.21)$$

where $r$ is called the gyrator coefficient.

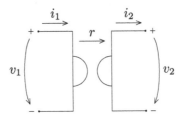

Figure 1.7  Diagram of an ideal gyrator.

## 1.3  MODELING OF MECHANICAL SYSTEMS

Mechanical systems can exhibit translational or rotational motions, or eventually a combination of both. In general, the governing equations of motion for mechanical systems in the 3-dimensional space require up to six variables, besides time, since a system is capable of three translational and three rotational motions. In this textbook, we consider only linear models and planar motions. Mechanical modeling is based on the concept

## 1.3.1 Translation Motion

The modeling of translational mechanical systems describes the relationship between two variables, namely the velocity $v$ (or, using a distinct notation, $\dot{x}$) and the force $f$, so that its scalar product corresponds to power $P$ (i.e., $v \cdot f = P$). The velocity and the force can be seen as an *across* or *effort* and a *through* or *flow* variable, respectively.

As mentioned for the electrical systems, from the mathematical point of view there is no significant difference if we consider the alternative assignment. Indeed, we find in the literature many works that adopt *through* or *flow* variable for $v$ and *across* or *effort* for $f$.

The velocity $v$ is related to the displacement $x$ by the equation:

$$v = \frac{\mathrm{d}x}{\mathrm{d}t}. \tag{1.22}$$

The relationship between the force $f$ and the momentum $p$ is given by:

$$f = \frac{\mathrm{d}p}{\mathrm{d}t}. \tag{1.23}$$

Alternatively, these equations can be written in the integral form yielding:

$$x = \int_0^t v(\tau)\mathrm{d}\tau, \tag{1.24}$$

and

$$p = \int_0^t f(\tau)\mathrm{d}\tau. \tag{1.25}$$

Mechanical systems obey by Newton's law stating that the algebraic sum of all forces equals zero, or by other words, the sum of applied forces must be equal to the sum of the reactive forces. Therefore, we can write:

$$\sum_{k=1}^{n} f_k(t) = 0, \tag{1.26}$$

where $n$ denotes the total number of forces acting on that point and $f_k$ stands for the $k$-th force.

The standard graphical representation of forces and velocities is designed by means of straight direction arrows.

In mechanical systems we can have two possible types of sources of energy, namely velocity and force. However, contrary to the case of electrical systems no graphical symbol other than a simple arrow is usually adopted.

We have three basic constitutive equations of the elements that relate the difference of velocities $v$ or of displacements $x$ between the two terminals, and the force $f$ through the elements. The basic elements are called the viscous friction, spring and mass that are described by:

$$f(t) = Bv(t), \qquad (1.27)$$

$$f(t) = Kx(t), \qquad (1.28)$$

$$f(t) = M\frac{\mathrm{d}v(t)}{\mathrm{d}t}, \qquad (1.29)$$

where $x(t) = \int_0^t v(\tau)\mathrm{d}\tau + x(0)$.

Friction forces are mostly non-linear in real-world and, therefore, Equation (1.27) is a modeling simplification. The symbol $B$ is called the damping coefficient. Equation (1.28) is called the Hooke's law and the symbol $K$ denotes the spring stiffness. Equation (1.29) is the Newton's second law and $M$ represents the inertia or mass.

The symbols for the three basic translational mechanical elements are depicted in Figure 1.8.

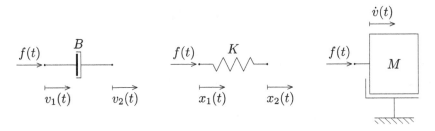

Figure 1.8 Basic translational mechanical elements: friction $B$, spring $K$ and mass $M$.

Expression (1.29) relates $v(t)$ and $f(t)$ in the differential form, but it can be written as:

$$v(t) = \frac{1}{M}p + v(0) = \frac{1}{M}\int_0^t f(\tau)d\tau + v(0), \qquad (1.30)$$

where $v(0)$ represents the velocity at $t = 0$, that is, the initial condition.

The friction dissipates energy (the so-called Joule effect $W = \frac{1}{2}Bv^2$), while the spring and the mass store energy in the potential and kinetic forms ($W = \frac{1}{2}Kx^2$ and $W = \frac{1}{2}Mv^2$), respectively. The units of the variables and elements are given in Table A.3.

The expressions (1.27)–(1.30) can be written in the Laplace domain. For zero initial conditions we have:

$$F(s) = BV(s), \qquad (1.31)$$

$$F(s) = KX(s) \Leftrightarrow F(s) = \frac{1}{sK}V(s), \qquad (1.32)$$

$$F(s) = sMV(s) \Leftrightarrow V(s) = \frac{1}{sM}F(s). \qquad (1.33)$$

Figure 1.9 shows an example of a translational mechanical system with an input force $f$ and one displacement $x$, and including the elements friction $B$, spring $K$ and mass $M$.

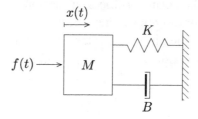

**Figure 1.9** Translational mechanical system with input force $f$ and displacement $x$, having the elements friction $B$, spring $K$ and mass $M$.

We can write the corresponding mathematical model:

$$f = Bv + M\frac{dv}{dt} + Kx. \qquad (1.34)$$

By assuming zero initial conditions we can obtain the transfer function between the input force and the output displacement:

$$\frac{X(s)}{F(s)} = \frac{1}{Ms^2 + Bs + K}. \tag{1.35}$$

Figure 1.10 shows a second example of a translational mechanical system with input force $f$, the elements friction $B$, spring $K$ and mass $M$, that exhibits two displacements $x_1$ and $x_2$.

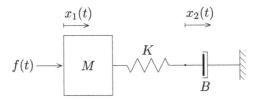

Figure 1.10 Translational mechanical system with input force $f$, including the elements friction $B$, spring $K$ and mass $M$, that exhibits two displacements $x_1$ and $x_2$.

In this case, we can write the mathematical model:

$$\begin{aligned} f &= M\ddot{x}_1 + K(x_1 - x_2), \\ K(x_1 - x_2) &= B\dot{x}_2. \end{aligned} \tag{1.36}$$

### 1.3.2 Rotational Motion

The modeling of rotational mechanical systems describes the relationship between two variables, namely the angular velocity $\omega$ (or, using a distinct notation, $\dot{\theta}$) and the torque $T$, so that its scalar product corresponds to power $P$ (i.e., $\omega \cdot T = P$). The angular velocity and the torque about a fixed axis can be seen as an *across* and a *through* variable, respectively. As before, we can consider the alternative, that is, assign the *through* or *flow* variable for $\omega$ and the *across* or *effort* for $T$.

The velocity $\omega$ is related to the angular displacement $\theta$ by the equation:

$$\omega = \frac{d\theta}{dt}. \tag{1.37}$$

The relationship between the torque $T$ and the rotational momentum $\mathcal{M}$ is given by:

$$f = \frac{d\mathcal{M}}{dt}. \tag{1.38}$$

Alternatively, these equations can be written in the integral form yielding:

$$\theta = \int_0^t \omega(\tau)d\tau, \quad (1.39)$$

and

$$\mathcal{M} = \int_0^t T(\tau)d\tau. \quad (1.40)$$

Mechanical systems obey Newton's law stating that the algebraic sum of all torques equals zero, or by other words, the sum of applied torques must be equal to the sum of the reactive torques. Therefore we can write:

$$\sum_{k=1}^n T_k(t) = 0, \quad (1.41)$$

where $n$ denotes the total number of torques acting on a fixed axis and $T_k$ stands for the $k$-th torque.

The standard graphical representation of torques and angular velocities is designed by means of curved arrows.

In rotational systems we can have two possible types of sources of energy, namely of angular velocity and torque. Usually, no graphical symbol other than a simple curved arrow is usually adopted.

We have three basic constitutive equations of the elements that relate the difference of angular velocities $\omega$ or of angular displacements $\theta$ across the two terminals, and the torque $T$ through the elements. The basic elements are called the rotational viscous friction, spring, and mass that are described by:

$$T(t) = B\omega(t), \quad (1.42)$$

$$T(t) = K\theta(t), \quad (1.43)$$

$$T(t) = J\frac{d\omega(t)}{dt}, \quad (1.44)$$

where $\theta(t) = \int_0^t \omega(\tau)d\tau + \theta(0)$.

The symbols $B$, $K$, and $J$ are called the rotational damping coefficient, spring stiffness, and mass or inertia, respectively. The symbols for the three basic rotational mechanical elements are depicted in Figure 1.11.

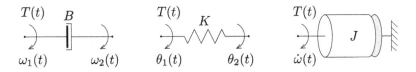

Figure 1.11 Basic rotational mechanical elements: friction $B$, spring $K$ and mass or inertia $J$.

Expression (1.44) relates $\omega(t)$ and $T(t)$ in the differential form, but it can be written as:

$$\omega(t) = \frac{1}{J}\mathcal{M} + \omega(0) = \frac{1}{J}\int_0^t T(\tau)d\tau + \omega(0), \qquad (1.45)$$

where $\omega(0)$ represents the angular velocity at $t=0$, that is, the initial condition.

The friction dissipates energy $W = \frac{1}{2}B\omega^2$, while the spring and the inertia store energy in the potential and kinetic forms ($W = \frac{1}{2}K\theta^2$ and $W = \frac{1}{2}J\omega^2$), respectively. The units of the variables and elements are given in Table A.4.

The expressions (1.42)–(1.45) can be written in the Laplace domain. For zero initial conditions we have:

$$T(s) = B\Omega(s), \qquad (1.46)$$

$$T(s) = K\Theta(s) \Leftrightarrow T(s) = \frac{1}{sK}\Omega(s), \qquad (1.47)$$

$$T(s) = sJ\Omega(s) \Leftrightarrow \Omega(s) = \frac{1}{sM}T(s). \qquad (1.48)$$

Figure 1.12 shows an example of a mechanical system with an input torque $T$ and one angular displacement $\theta$, and including the elements friction $B$, spring $K$ and inertia $J$.

We can write the corresponding mathematical model:

$$T = B\omega + J\frac{d\omega}{dt} + K\theta. \qquad (1.49)$$

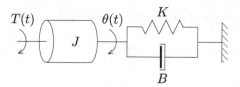

**Figure 1.12** Rotational mechanical system with input torque $T$ and angular displacement $\theta$, having the elements friction $B$, spring $K$ and inertia $J$.

By assuming zero initial conditions we can obtain the transfer function between the input torque and the output angular displacement:

$$\frac{\Theta(s)}{T(s)} = \frac{1}{Js^2 + Bs + K}. \tag{1.50}$$

Figure 1.13 shows a second example of a mechanical system with input torque $T$, elements friction $B$, spring $K$ and inertia $J$, and exhibiting two displacements $\theta_1$ and $\theta_2$.

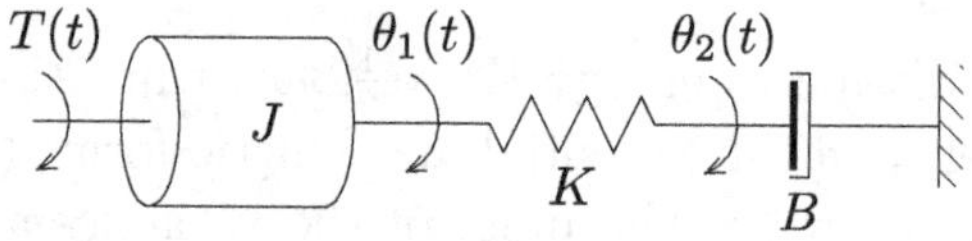

**Figure 1.13** Rotational mechanical system with input torque $T$, including the elements friction $B$, spring $K$ and inertia $J$, and exhibiting two displacements $\theta_1$ and $\theta_2$.

In this case, we can write the mathematical model:

$$\begin{aligned} T &= J\ddot{\theta}_1 + K\left(\theta_1 - \theta_2\right), \\ K\left(\theta_1 - \theta_2\right) &= B\dot{\theta}_2. \end{aligned} \tag{1.51}$$

## 1.4 TRANSMISSION SYSTEMS

Electro-mechanical systems often include mechanical transmissions between the motor and the load axes. In the next sub-sections we shall find the simplified linear models of the most frequent.

### 1.4.1 Conversion between Rotational Motions

An ideal rotational mechanical transformer relates the angular velocity and torque at the primary (also called port 1) to the angular velocity and torque at the secondary (or port 2) by means of a linear relation. We can consider a gear train or a belt and pulley system represented schematically in Figure 1.14.

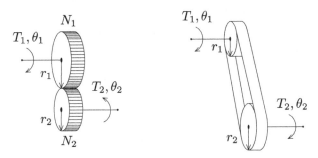

Figure 1.14 Rotational to rotational motion transformation: gear train and belt and pulley system.

The relationships between torques $T_1$ and $T_2$, angular displacement $\theta_1$ and $\theta_2$, teeth numbers $N_1$ and $N_2$, and radii $r_1$ and $r_2$ of the gear train are derived from the following ideas:

- The number of teeth on the surface of the gears is proportional to the radii of the gears:

$$\frac{r_1}{r_2} = \frac{N_1}{N_2} \tag{1.52}$$

- The distance traveled along the surface of each gear is identical (i.e., there is no slip):

$$\theta_1 \cdot r_1 = \theta_2 \cdot r_2 \tag{1.53}$$

- The work done by one gear is equal to that of the other (the gear is assumed ideal with no losses):

$$T_1 \cdot \theta_1 = T_2 \cdot \theta_2. \tag{1.54}$$

Joining the equations (1.52)–(1.54) we can write:

$$\frac{T_1}{T_2} = \frac{\theta_2}{\theta_1} = \frac{\omega_2}{\omega_1} = \frac{r_1}{r_2} = \frac{N_1}{N_2} = n. \qquad (1.55)$$

The gear and belt and pulley systems have different constructions, but from the ideal point of view they are described by the same models.

Let us consider the motor-load assembly, shown in Figure 1.15, with rigid shafts and one gear train with transformation ratio $\frac{r_1}{r_2} = \frac{N_1}{N_2} = n$. The symbols $J_1$ and $J_2$ denote the inertia at the primary and secondary of the gear. Moreover, let $B_1$ and $B_2$ stand for the friction coefficients at the two sides, and let $T_m$ and $T_L$ represent the torques provided by the motor and delivered to the load, respectively.

After using (1.55) we obtain the model:

$$T_m = \left[J_1 + J_2\left(\frac{N_1}{N_2}\right)^2\right]\ddot{\theta}_1 + \left[B_1 + B_2\left(\frac{N_1}{N_2}\right)^2\right]\dot{\theta}_1 + \left(\frac{N_1}{N_2}\right)T_L. \qquad (1.56)$$

We conclude that if we use a gear train, then the equivalent inertia and friction coefficient of the motor and load are given by $J_{eq} = J_1 + J_2 n^2$ and $B_{eq} = B_1 + B_2 n^2$, respectively, while the load torque seen by the motor is $T_{eq} = nT_L$.

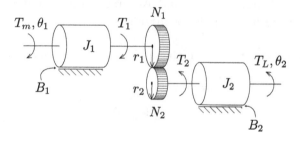

Figure 1.15  Gear train with friction and inertia.

### 1.4.2  Conversion from Rotational to Translational Motion

In motion control systems, it is often necessary to convert from rotational to translational motion. Figure 1.16 shows three systems for implementing the rotational to translational motion transformation, consisting of the lead screw, rack and pinion, and belt and pulley mechanisms.

For the lead screw we can write:

$$\frac{\theta}{2\pi} = \frac{x}{h} \qquad (1.57)$$

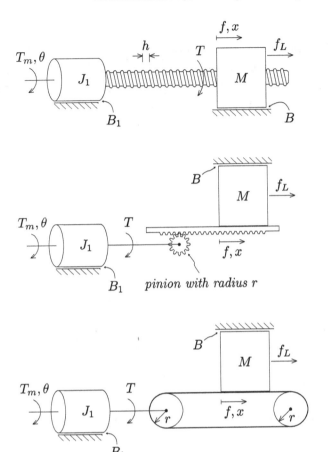

Figure 1.16 Rotational to translational motion transformation: lead screw, rack and pinion, and belt and pulley mechanisms.

and

$$f \cdot x = T \cdot \theta, \tag{1.58}$$

where $h$ represents the screw pitch. Therefore, for a motor delivering a torque $T_m$ and load with inertia $M$, friction $B$ and requiring a force $f_L$ we obtain:

$$T_m = \left[J_1 + M\left(\frac{h}{2\pi}\right)^2\right]\ddot{\theta} + \left[B_1 + B\left(\frac{h}{2\pi}\right)^2\right]\dot{\theta} + \left(\frac{h}{2\pi}\right)f_L. \tag{1.59}$$

We conclude that if we use a lead screw, then the equivalent inertia and friction coefficient of the motor and load are given by $J_{eq} = J_1 + Mn^2$ and $B_{eq} = B_1 + B_2 n^2$, respectively, while the load torque seen by the motor is $T_{eq} = n f_L$, with $n = \frac{h}{2\pi}$ representing the transformation ratio.

For the rack and pinion system we can write:

$$r \cdot \theta = x \tag{1.60}$$

and

$$f \cdot x = T \cdot \theta, \tag{1.61}$$

where $r$ represents the radius. Therefore, for a motor delivering a torque $T_m$ and load with inertia $M$, friction $B$ and requiring a force $f_L$ we obtain:

$$T_m = \left(J_1 + Mr^2\right) \ddot{\theta} + \left(B_1 + Br^2\right) \dot{\theta} + r f_L. \tag{1.62}$$

In summary, if we use a rack and pinion system, then the equivalent inertia and friction coefficient of the motor and load are given by $J_{eq} = J_1 + Mr^2$ and $B_{eq} = B_1 + Br^2$, respectively, while the load torque seen by the motor is $T_{eq} = r f_L$, with $r$ representing the transformation ratio.

For the belt and pulley mechanism we have exactly the same expressions:

$$T_m = \left(J_1 + Mr^2\right) \ddot{\theta} + \left(B_1 + Br^2\right) \dot{\theta} + r f_L. \tag{1.63}$$

In conclusion, if we use a belt and pulley system, then the equivalent inertia and friction coefficient of the motor and load are given by $J_{eq} = J_1 + Mr^2$ and $B_{eq} = B_1 + Br^2$, respectively, while the load torque seen by the motor is $T_{eq} = r f_L$, with $r$ representing the transformation ratio.

### 1.4.3 Translational to Translational Motion Transformation

Mechanical levers (Figure 1.17) are rarely used since they require almost static operating conditions.

For very small values of the angle $\alpha$ we can say that the arcs $A_1$ and $A_2$ are close to the linear displacements $x_1$ and $x_2$, respectively, and therefore:

$$\frac{f_1}{f_2} = \frac{x_2}{x_1} = \frac{l_2}{l_1}. \tag{1.64}$$

Mathematical Models of Dynamical Systems ■ 19

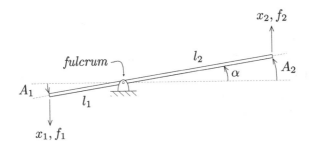

Figure 1.17 Translational to translational motion transformation: mechanical lever.

### 1.4.4 Permanent Magnet DC Motors

In this sub-section, we consider the permanent magnet DC motor with the circuit diagram represented in Figure 1.18. The armature is modeled as a circuit with resistance $R_a$ in series with an inductance $L_a$. When the rotor rotates a back electromotive force $e_b$ emerges in the armature circuit.

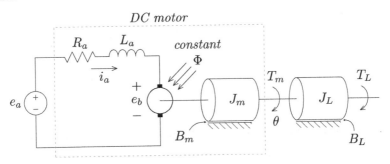

Figure 1.18 Permanent magnet DC motor.

For the permanent magnet DC motor the model can be written as follows:

$$\begin{aligned} T_m(t) &= K_i i_a(t) \\ e_{b(t)} &= K_b \dot{\theta}(t) \\ e_a(t) &= R_a i_a(t) + L_a \frac{di_a(t)}{dt} + e_b(t) \\ T_m(t) &= (J_m + J_L)\ddot{\theta}(t) + (B_m + B_L)\dot{\theta}(t) + T_L(t), \end{aligned} \quad (1.65)$$

where $K_i$ and $K_b$ are motor constants, $\theta$ is the motor and load angular displacement, $T_m$ and $T_L$ represent the motor and load torques. Additionally, $J_m$ and $J_L$ are the motor and load intertias, respectively, and $B_m$ and $B_L$ denote the corresponding angular viscous friction coefficients.

When using units in the international system and considering an ideal motor, the principle of conservation of energy, expressed as $P = T_m \dot{\theta} = E_b i_a$, that is, identical power in the electrical and in the mechanical domains, implies that $K_i = K_b$.

Applying the Laplace transform we obtain:

$$\begin{aligned} T_m(s) &= K_i I_a(s) \\ E_b(s) &= s K_b \Theta(s) \\ E_a(s) &= (R_a + s L_a) I_a(s) + E_b(s) \\ T_m(s) &= \left[ s^2 (J_m + J_L) + s (B_m + B_L) \right] \Theta(s) + T_L(s). \end{aligned} \quad (1.66)$$

The corresponding block diagram is depicted in Figure 1.19.

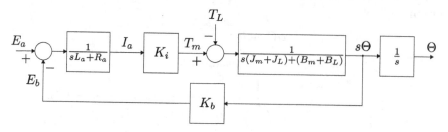

Figure 1.19 Block diagram of the permanent magnet DC motor.

## 1.5 MODELING OF HYDRAULIC SYSTEMS

In this sub-section we focus on incompressible fluid systems, so that the fluid volume remains constant. Like for electrical systems we have two variables, the pressure drop $p$ *across* the element endpoints and the volumetric flow rate $q$ *through* the element. The scalar product of these two variables corresponds to power $P$ (i.e., $p \cdot q = P$).

The pressure $p$ is related to the pressure momentum $\Gamma$ by:

$$p = \frac{d\Gamma}{dt}. \quad (1.67)$$

The volumetric flow rate $q$ is related to the fluid volume $V$ by the equation:

$$q = \frac{dV}{dt}. \tag{1.68}$$

Alternatively, these equations can be written in the integral form so that:

$$\Gamma = \int_0^t p(\tau)d\tau, \tag{1.69}$$

and

$$V = \int_0^t q(\tau)d\tau. \tag{1.70}$$

Also, such as for electrical systems, we have three one-port passive elements, namely the fluid resistance $R$, inertance $L$ and reservoir $C$, as represented in Figure 1.20.

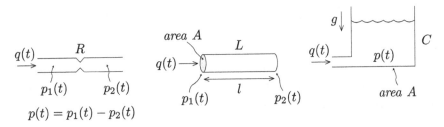

Figure 1.20 Basic hydraulic elements: fluid resistance $R$, inertance $L$ and capacitance $C$.

The dissipative element resistance $R$ relates the pressure drop $p$ across the endpoints of a pipe to the volumetric flow rate $q$. Depending on the type of flow (i.e., laminar or turbulent) the fluid resistance relationship can be linear or nonlinear. For a laminar flow, we define

$$p(t) = Rq(t). \tag{1.71}$$

The mechanism for storing effort energy consists of the kinetic energy associated with the moving body of fluid molecules and can be modeled by the inertance $L$ with the constitutive equation

$$p(t) = L\frac{dq(t)}{dt}, \tag{1.72}$$

where $L = \frac{\rho l}{A}$, with $\rho$ denoting the mass density of the fluid, and $A$ and $l$ the cross-sectional area and length of the pipe.

An open tank or reservoir which is fed with working fluid is a store of fluid flow energy. The corresponding constitutive equation for the fluid capacitance $C$ is given by:

$$q(t) = C \frac{dp(t)}{dt}, \tag{1.73}$$

where $C = \frac{A}{\rho g}$, with $\rho$ denoting the mass density of the fluid, $A$ the cross-sectional area of the reservoir and $g$ the acceleration of gravity.

Expressions (1.72) and (1.73) relate $p(t)$ and $q(t)$ in the differential form, but we can write them in the integral form:

$$q(t) = \frac{1}{L}\Gamma + q(0) = \frac{1}{L}\int_0^t p(\tau)d\tau + q(0), \tag{1.74}$$

$$p(t) = \frac{1}{C}V + p(0) = \frac{1}{C}\int_0^t q(\tau)d\tau + p(0), \tag{1.75}$$

where $q(0)$ and $p(0)$ represent the initial conditions, namely, the volumetric flow rate and pressure at $t = 0$.

The expressions (1.72)–(1.75) can be written in the Laplace domain. For zero initial conditions we have:

$$P(s) = RQ(s), \tag{1.76}$$

$$P(s) = sLQ(s) \Leftrightarrow Q(s) = \frac{1}{sL}P(s), \tag{1.77}$$

$$Q(s) = sCP(s) \Leftrightarrow P(s) = \frac{1}{sC}Q(s). \tag{1.78}$$

Figure 1.21 shows an example of a hydraulic system with input and output pressures $p_i$ and $p_o$, respectively, with a fluid resistance $R$, inertance $L$ and capacitance $C$ in series.

We can write the mathematical model:

$$p_i = Rq + L\frac{dq}{dt} + \frac{1}{C}\int_0^t q(\tau)d\tau + p(0), \tag{1.79}$$

$$p_o = \frac{1}{C}\int_0^t q(\tau)d\tau + p(0), \tag{1.80}$$

where $q$ is the volumetric flow rate flowing through the three elements.

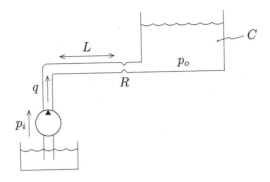

Figure 1.21 Hydraulic system: fluid resistance $R$, inertance $L$ and capacitance $C$ in series.

By assuming zero initial conditions, we can obtain the transfer function, yielding:

$$\frac{P_o(s)}{P_i(s)} = \frac{1}{LCs^2 + RCs + 1}. \qquad (1.81)$$

Figure 1.22 shows an example of a hydraulic system with input flow $q_i$ and a fluid resistance $R$, inertance $L$ and capacitance $C$ in parallel.

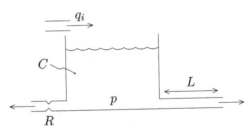

Figure 1.22 Hydraulic system: fluid resistance $R$, inertance $L$ and capacitance $C$ in parallel.

We can write the mathematical model:

$$q_i = \frac{1}{R}p + \frac{1}{L}\int_0^t p(\tau)\mathrm{d}\tau + C\frac{\mathrm{d}p}{\mathrm{d}t} + q(0). \qquad (1.82)$$

Let us now consider the two-port fluid transformer represented on Figure 1.23 with pressures and volumetric flow rates $p_1$ and $q_1$ in the primary and $p_2$ and $q_2$ in the secondary circuit.

Let us also consider that the displacement $x$ is identical in both pistons with cross-areas $A_1$ and $A_2$ in the primary and secondary, respectively. We obtain $A_1 x = V_1 \Leftrightarrow A_1 \dot{x} = q_1$, $A_2 x = V_2 \Leftrightarrow A_2 \dot{x} = q_2$ and considering an ideal device we have identical power at input and output $p_1 \cdot q_1 = p_2 \cdot q_2$. Therefore, we can write the equations:

$$\frac{p_1}{p_2} = \frac{q_2}{q_1} = \frac{1}{n}, \qquad (1.83)$$

where $n = \frac{A_1}{A_2}$ is called the transformation ratio.

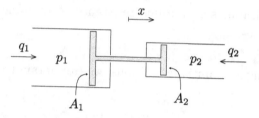

Figure 1.23  Diagram of an ideal fluid transformer.

A simple fluid to mechanical transformer is depicted in Figure 1.24. This two-port element includes a piston with area $A$ connected to a rigid axis.

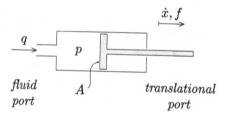

Figure 1.24  Diagram of an ideal fluid to mechanical transformer.

In the fluid port we have pressure $p$ and flow rate $q$, while in the mechanical translational port we have velocity $\dot{x}$ and force $f$.

Under ideal circumstances we have the model:

$$pA = f$$
$$\dot{x}A = q. \qquad (1.84)$$

The hydraulic mechanism of Figure 1.25 represents the fundamental design of an hydraulic press. The two-port system includes a fluid interconnecting two pistons of areas $A_1$ and $A_2$. The pistons have displacements $x_1$ and $x_2$ under the action of forces $f_1$ and $f_2$, respectively.

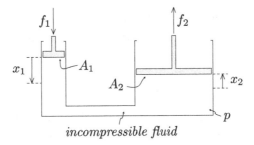

Figure 1.25  Diagram of a hydraulic press.

Then, we can write:

$$pA_1 = f_1$$
$$pA_2 = f_2 \qquad (1.85)$$
$$\dot{x}_1 A_1 = \dot{x}_2 A_2,$$

where $p$ stands for the pressure in the fluid supposed to be incompressible. If we consider the ratio $n = \frac{A_1}{A_2}$ then we have the transformer equation:

$$\frac{f_1}{f_2} = \frac{\dot{x}_2}{\dot{x}_1} = n. \qquad (1.86)$$

## 1.6  MODELING OF THERMAL SYSTEMS

The two key variables in a thermal process are temperature $T$ and heat flow rate $q$, corresponding to the *across* and *through* variables, respectively. In this case the product of the two variables is energy.

The heat flow rate $q$ is related to the total heat $H$ by:

$$q = \frac{dH}{dt}, \qquad (1.87)$$

or, alternatively in the integral form:

$$H = \int_0^t q(\tau)d\tau. \qquad (1.88)$$

Heat transfer systems can be modeled by the one-port elements thermal resistance $R$ and capacitance $C$, that behave analogously to those mentioned in electrical systems. Thermal resistance is a property of each material and is modeled by the dissipative element resistance $R$ that relates the temperature drop $T$ across the endpoints of a object to the heat flow rate $q$, so that

$$T(t) = Rq(t). \tag{1.89}$$

The ability of a given material to store heat is a measure of its thermal capacity. The corresponding constitutive equation for the thermal capacitance $C$ is given in the differential form by the equation:

$$q(t) = C\frac{\mathrm{d}T(t)}{\mathrm{d}t}. \tag{1.90}$$

In the integral form we can write:

$$T(t) = \frac{1}{C}H + T(0) = \frac{1}{C}\int_0^t q(\tau)\mathrm{d}\tau + T(0). \tag{1.91}$$

Apparently there is not a thermal element displaying an energy storage mechanism of $T$ and, therefore, thermal systems do not include the inductance element.

Figure 1.26 shows an example of a thermal system with input and output temperatures $T_1$ and $T_2$, respectively, with one thermal resistance $R$, and one capacitance $C$.

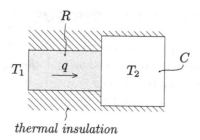

Figure 1.26 Thermal system with one thermal resistance $R$ and capacitance $C$.

We can write the mathematical model:

$$T_1 - T_2 = Rq,$$
$$T_2 = C\frac{\mathrm{d}q}{\mathrm{d}t}. \tag{1.92}$$

CHAPTER 2

# Bond Graph Modeling

## 2.1 INTRODUCTION

A bond graph is a graphical approach for the modeling of physical systems. A bond graph has some similarities to other system representations, such as block diagrams and signal-flow graphs. However, in the bond graphs the arcs/lines connecting elements stand for a *bi-directional exchange of power*, while those adopted in block diagrams and signal-flow graphs correspond merely to a *uni-directional flow of information*. As a consequence, a bond graph is in general a more compact and 'tidy' chart than the other system representations. Bond graphs are multi-energy domain, that is, they can incorporate multiple domains namely mechanical, electrical, hydraulic, thermal and others.

Bond graphs were designed in 1959 by Henry Paynter [1]. He proposed the idea of portraying systems in terms of power bonds. The bond graph formalism was further developed by several researchers and we can mention the works of Karnopp, Rosenberg, Thoma, Breedveld and Borutzky just to name a few [2, 3, 4, 5, 6]. More recently, several researchers extended the bond graph modeling technique to hydraulics, mechatronics, and non-energetic systems like architecture [7, 8, 9], economics [10, 11, 12, 13, 14, 15], fractional calculus [16, 17] and others [18, 19, 20, 21, 22].

## 2.2 BASIC CONCEPTS

The bond graphs are composed of the 'bonds' that link together elements. Each bond represents the instantaneous flow of power or energy. The connection between the elements are accomplished by means of 'junction' structures that are a manifestation of the system constraints.

The basic bond graph variables form the 'tetrahedron of state', as represented in Figure 2.1, and consist of:

- effort $e(t)$
- flow $f(t)$
- time integral of effort $e_a(t) = \int_0^t e(\tau)\mathrm{d}\tau + e_a(0)$
- time integral of flow $f_a(t) = \int_0^t f(\tau)\mathrm{d}\tau + f_a(0)$.

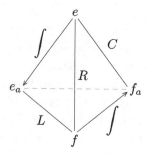

Figure 2.1 'Tetrahedron of state' by Henry Paynter.

Often, the accumulated effort and flow, $e_a(t)$ and $f_a(t)$, are denoted by impulse and charge, $p(t)$ and $q(t)$, respectively. The fundamental idea of a bond graph is that power $P$ is transmitted between connected components by a combination of the generalized variables 'effort' and 'flow', $e(t)$ and $f(t)$, as represented in Figure 2.2, where the symbols $\mathcal{E}_1$ and $\mathcal{E}_2$ stand for two systems elements and the power is $P(t) = e(t) \cdot f(t)$. A half-arrow ⟶ is the sign convention for power bond, with the figure corresponding to the case where element $\mathcal{E}_1$ is delivering power to element $\mathcal{E}_2$. A full arrow --→ is used to indicate signal, or information, connection, and in this case the amount of power flowing through the harrowed arc/line is zero.

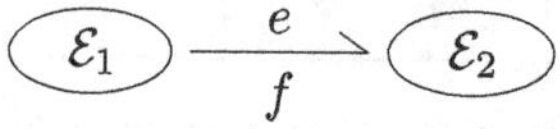

Figure 2.2 Bond graph sign convention: A half-arrow for power bond (in the case where element $\mathcal{E}_1$ is delivering power to element $\mathcal{E}_2$).

Table 2.1 lists common assignments of effort and flow in five physical domains. Nonetheless, the definition of the $e$ and $f$ variables is arbitrary and the reverse pair can also be adopted. Therefore, for some systems different authors assign distinct $e$ and $f$ variables, e.g., [4] and [23, 24]. The assignments marked with asterisk are used only in a limited number of examples in this book, being called as 'Option 2'. Therefore we have:

- Option 1: {linear/angular velocity, force/torque}→{effort, flow}
- Option 2: {force/torque, linear/angular velocity}→{effort, flow}

**TABLE 2.1** Common assignments of effort and flow in five physical domains. Those marked with asterisk are used only a limited number of examples in this book, and are called as 'Option 2'.

| System | Effort ($e$) | Flow ($f$) |
|---|---|---|
| Electrical | Voltage ($v$) | Current ($i$) |
| Mechanical (lin) | Velocity ($v$) | Force ($f$) |
| * Mechanical (lin) | Force ($f$) | Velocity ($v$) |
| Mechanical (rot) | Angular velocity ($\omega$) | Torque ($T$) |
| * Mechanical (rot) | Torque ($T$) | Angular velocity ($\omega$) |
| Hydraulic | Pressure ($p$) | Volume flow rate ($q = \frac{dV}{dt}$) |
| Thermal | Temperature ($T$) | Heat flow rate ($q$) |

Thermal systems can be thought as driven by temperature ($T$) and heat flow rate ($q$), although the product of the two variables is energy (i.e., $W = T \cdot q$), not power as in the remaining sets of system variables.

As an historical note, it is interesting to note that Henry Paynter considered not of special relevance the relationship between $e_a(t)$ and $f_a(t)$ (left side of Figure 2.1), but that issue was later resumed by Leon Chua [25, 26, 27] in the scope of the memristor concept (right side of Figure 2.1).

The concept of memristor and further generalizations, such as meminductor ($L_M$) and memcapacitor ($C_M$), or simply denoted as high-order elements, are often represented as in Figure 2.3, but their discussion is outside the scope of this book. Interested readers can refer to [25, 26, 27, 28, 29, 30, 31].

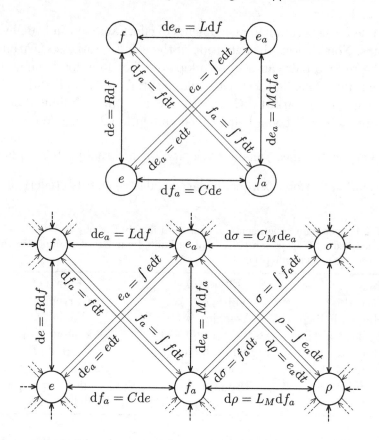

Figure 2.3  The concept of memristor denoted by $M$ (top) by Leon Chua and the high-order elements (bottom) with meminductor and memcapacitor denoted by $L_M$ and $C_M$, respectively.

## 2.3  BOND GRAPH ELEMENTS

In the bond graph formalism three groups of basic objects are adopted:

- One port elements:
  - energy sources
  - energy stores
  - dissipators
- Two port elements
- Junctions

### 2.3.1 One-Port Elements: Energy Sources

We have two basic energy sources, one for each of the bond graph variable pair required to specify power. The effort and flow sources are denoted by $SE$ and $SF$, respectively. Therefore, the abstract generalized effort and flow sources produce $e$ and $f$ to the interconnected element $\mathcal{E}$, respectively, as represented in Figure 2.4. For example, electrical systems have sources of voltage and current. Table 2.2 lists the energy sources for the generalized and five energy domains.

$$SE_{e} \xrightarrow{e} \mathcal{E}$$

$$SF_{f} \xrightarrow{f} \mathcal{E}$$

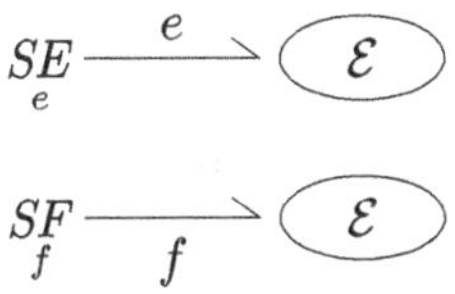

Figure 2.4 Generalized sources of effort ($SE$) and flow ($SF$).

TABLE 2.2 Energy sources for the generalized and five energy domains.

| Energy domain | Effort source | Flow source |
|---|---|---|
| Generalized | Effort $SE_{e} \longrightarrow$ | Flow $SF_{f} \longrightarrow$ |
| Electrical | Voltage $SE_{v} \longrightarrow$ | Current $SF_{i} \longrightarrow$ |
| Mechanical translational | Velocity $SE_{v} \longrightarrow$ | Force $SF_{f} \longrightarrow$ |
| Mechanical rotational | Angular velocity $SE_{\omega} \longrightarrow$ | Torque $SF_{T} \longrightarrow$ |
| Hydraulic | Pressure $SE_{p} \longrightarrow$ | Fluid flow $SF_{q} \longrightarrow$ |
| Thermal | Temperature $SE_{T} \longrightarrow$ | Heat flow $SF_{q} \longrightarrow$ |

## 2.3.2 One-Port Elements: Energy Stores

In general we have two energy stores, one for effort and another for flow store, with the exception of thermal systems where we have only the flow store.

The first type of device stores energy as the time integral of the effort variable (i.e., $p$) applied to its port. The flow device is then given by its material properties:

- In general, the flow is given by a non-linear function of the accumulation of effort: $f = \varphi_L^{-1}(p)$
- In the linear case, the effort is given by: $f = L^{-1}p$

For example, the electrical component is the inductor $L$ and $p$ is magnetic flux. Table 2.3 lists the effort stores for the generalized and five energy domains.

The second type of device stores energy as the time integral of the flow variable (i.e., $q$) applied to its port. The effort device is then given by its material properties:

- In general, the effort is given by a non-linear function of the accumulation of flow: $e = \varphi_C^{-1}(q)$
- In the linear case, the effort is given by: $e = C^{-1}q$

For example, the electrical component is the capacitor $C$ and $q$ is charge. Table 2.4 lists the effort stores for the generalized and five energy domains.

TABLE 2.3 Effort stores for the generalized and five energy domains, with: $\lambda$ magnetic flux, $\Gamma$ fluid momentum, $x$ translational displacement, $\theta$ rotational displacement.

| Element | Dynamic relation | Linear constructive relation |
|---|---|---|
| Generalized | $e_a = \int e\,dt$ | $f = \frac{1}{L}e_a$ |
| Electrical (inductor) | $\lambda = \int v\,dt$ | $i = \frac{1}{L}\lambda$ |
| Translational mechanical (compliance) | $x = \int v\,dt$ | $f = Kx$ |
| Rotational mechanical (compliance) | $\theta = \int \omega\,dt$ | $T = K\theta$ |
| Hydraulic (inertance) | $\Gamma = \int p\,dt$ | $q = \frac{1}{L}\Gamma$ |
| Thermal | | no equivalent |

TABLE 2.4 Flow stores for the generalized and five energy domains, with: $q$ electrical charge, $p$ translational momentum, $\mathcal{M}$ rotational momentum, $V$ fluid volume, $H$ total heat.

| Element | Dynamic relation | Linear constructive relation |
|---|---|---|
| Generalized | $f_a = \int f \mathrm{d}t$ | $e = \frac{1}{C} f_a$ |
| Electrical (capacitor) | $q = \int i \mathrm{d}t$ | $v = \frac{1}{C} q$ |
| Translational mechanical (inertia) | $p = \int f \mathrm{d}t$ | $v = \frac{1}{M} p$ |
| Rotational mechanical (inertia) | $\mathcal{M} = \int T \mathrm{d}t$ | $\omega = \frac{1}{J} \mathcal{M}$ |
| Hydraulic (fluid reservoir) | $V = \int q \mathrm{d}t$ | $p = \frac{1}{C} V$ |
| Thermal (thermal capacity) | $H = \int q \mathrm{d}t$ | $T = \frac{1}{C} H$ |

### 2.3.3 One-Port Elements: Dissipators

In what concerns energy dissipation a single element type is required to model the basic phenomenon. In this case we have a device whose effort and flow variables are statically constrained, and we can write:

- In the non-linear case: $e = \varphi(f)$
- In the linear case: $e = Rf$

For example, the electrical component is the resistor $R$. Table 2.5 lists the energy dissipators for the generalized and five energy domains.

TABLE 2.5 Energy dissipators for the generalized and five energy domains.

| Element | Linear constructive relation |
|---|---|
| Generalized | $e = Rf$ |
| Electrical (resistor) | $v = Ri$ |
| Translational mechanical (friction) | $f = Bv$ |
| Rotational mechanical (friction) | $T = B\omega$ |
| Hydraulic (fluid dissipator) | $p = Rq$ |
| Thermal (Fourier law) | $T = Rq$ |

## 2.3.4 Two-Port Elements

Two-ports elements are power conserving devices and they can be of two types:

- Transformer ($TF$) described by the equations $e_1(t) = n^{-1} \cdot e_2(t)$ and $f_1(t) = n \cdot f_2(t)$, where $n$ is the transformer ratio (units: dimensionless)

- Gyrator ($GY$) described by the equations $e_1(t) = r \cdot f_2(t)$ and $f_1(t) = r^{-1} \cdot e_2(t)$, where $r$ is the gyration resistance (units: $\Omega$)

Therefore, the expression $P = e_1 \cdot f_1 = e_2 \cdot f_2$ verifies the conservation of power. The standard and the bond graph symbols are represented in Figures 2.5 and 2.6, respectively.

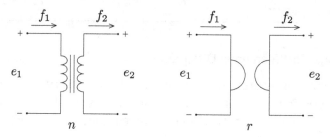

Figure 2.5 The standard symbols of the $TF$ (left) and $GY$ (right).

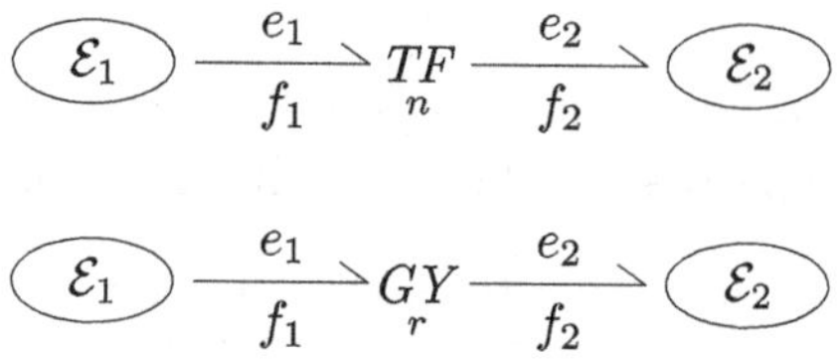

Figure 2.6 The bond graph symbols of the $TF$ and $GY$ (in the case where element $\mathcal{E}_1$ is delivering power to element $\mathcal{E}_2$).

Physical examples of $TF$ are the electrical transformer, mechanical gear trains and levers. For the $GY$, the gyroscope is a familiar example. Some authors consider that the $TF$ is less general than the $GY$. In fact, the association of two $TF$ with transform ratios $n_1$ and $n_2$ in series yields only a $TF$ with transform ratio $n = n_1 \cdot n_2$. On the other hand, the association of two $GY$ with ratios $r_1$ and $r_2$ in series (see next sub-section for junctions) is equivalent to one $TF$ with transform ratio $n = \frac{r_2}{r_1}$.

As mentioned initially, the assignment of the $e$ and $f$ variables is arbitrary. For example, the hydraulic piston (see Figure 2.7) can be represented by a $TF$, or alternatively a $GY$, by switching the $e$ and $f$ variables in one of the ports as represented in Figures 2.8-2.9.

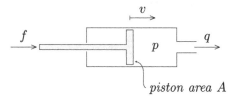

Figure 2.7  The hydraulic piston.

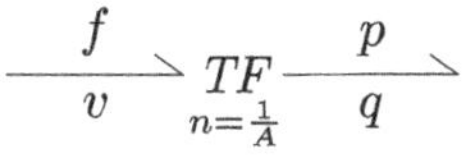

$$\xrightarrow{f}_{v} TF_{n=\frac{1}{A}} \xrightarrow{p}_{q}$$

{force, velocity}→{effort, flow}

Figure 2.8  The bond graph representation of the hydraulic piston as a $TF$ when assigning the effort and flow, $e$ and $f$, for the variables force and velocity, $f$ and $v$.

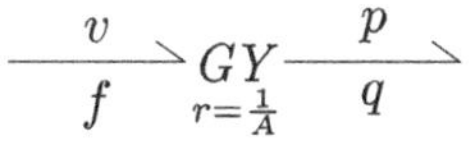

$$\xrightarrow{v}_{f} GY_{r=\frac{1}{A}} \xrightarrow{p}_{q}$$

{velocity, force}→{effort, flow}

Figure 2.9  The bond graph representation of the hydraulic piston as a $GY$ when assigning the effort and flow, $e$ and $f$, for the variables velocity and force, $v$ and $f$.

### 2.3.5 Junctions

We have two fundamental inter-connective constraints modeled in the bond graph formalism, namely the:

- compatibility constraint on the effort variables, denoted as a 1 junction or effort junction

- continuity constraint on the flow variables, denoted as a 0 junction, or flow junction

The 1 junction (or effort junction) imposes equal flow in the inter-connected elements, while the efforts sum up to zero (assuming the same power orientation). This constraint is modeled in bond graph by an $n$-port device known as effort junction: $f_1 = f_2 = \cdots = f_n$ and $e_1 + e_2 + \cdots + e_n = 0$ as represented schematically in Figure 2.10.

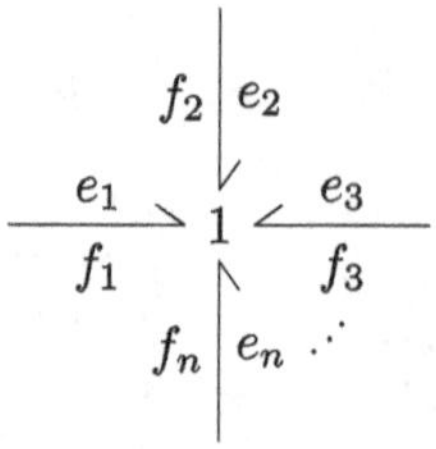

Figure 2.10  Schematic bond graph representation of a 1 junction.

For example, in electrical systems the constraint is a general statement of the Kirchhoff's Voltage Law. Figure 2.11 shows four examples of a 1 junction in the cases of electrical, mechanical, hydraulic and thermal systems.

The 0 junction (or flow junction) imposes identical efforts in the inter-connected elements, while the flows sum up to zero (assuming the same power orientation). This constraint is modeled in bond graph by an $n$-port device known as flow junction: $e_1 = e_2 = \cdots = e_n$ and $f_1 + f_2 + \cdots + f_n = 0$ as represented schematically in Figure 2.12.

For example, in electrical systems the constraint is a general statement of the Kirchhoff's Current Law. Figure 2.13 shows four examples of a 0 junction in the cases of electrical, mechanical, hydraulic and thermal systems.

### 2.3.6  Modulated Elements

Often the properties of an element depend on an external input or system variable. Controlling the output of a source is the most common modeling procedure in which inputs are applied to a system representation. Virtually, all bond graph elements can be considered and, therefore, we have the following cases that depend on an additional control variable $u$:

- Controlled sources, of two types:
  - modulated effort source, *MSE*
  - modulated flow source, *MSF*

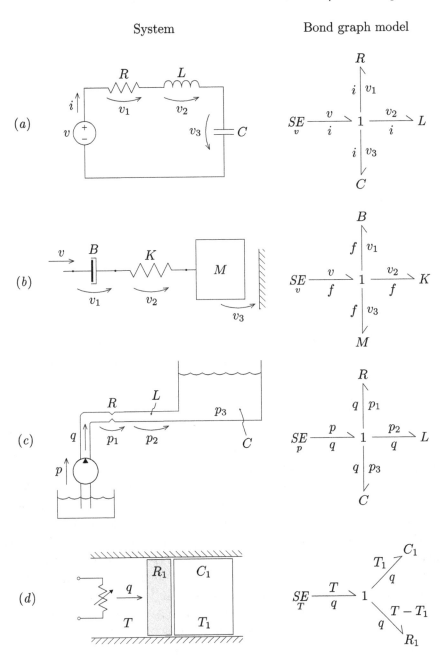

Figure 2.11 Simple examples of a 1 junction for (a) electrical, (b) mechanical, (c) hydraulic and (d) thermal systems.

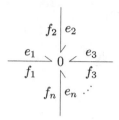

Figure 2.12  Schematic bond graph representation of a 0 junction.

- Modulated One-ports, of three types:
  - modulated effort store, *ML* with value $L(u)$
  - modulated flow store, *MC*, with value $C(u)$
  - modulated resistor, *MR*, with value $R(u)$

- Modulated Two-ports, of two types:
  - modulated transformer, *MTF*, with transformation ratio $n(u)$
  - modulated gyrator, *MGY*, with gyrator coefficient $r(u)$.

For example, Figure 2.14 depicts the bond graph symbols for the *MTF* and the *MGY*.

We should note that the symbol adopted here for the modulated resistor, *MR*, should not be confused with the symbol often used for the 'memristor' mentioned in Section 2.2.

Figures 2.15 and 2.16 show examples of modulated elements in electromechanical and electronic systems, respectively. The first circuit consists of an ideal DC servomotor with separate fixed field current (or, excitation) with an applied voltage $v$ to and requiring the current $i$. We can write the linear model $T = (Ki_F)i$ and $v = (Ki_F)\omega$, where $i_F$ is the field current and $K$ is a constant that characterizes the motor. If we assign the angular velocity $\omega$ and the torque $T$ as effort and flow, respectively, then we have a *MTF* with transformation ratio $n(i_F) = (Ki_F)^{-1}$. Inversely, if we assign the torque $T$ and angular velocity $\omega$ as effort and flow, respectively, then we have a *MGY* with gyration ratio $r(i_F) = Ki_F$. The second circuit corresponds to a transistor model, namely to an equivalent circuit for common emitter. In this case we have a *MSF* that yields a current proportional to the voltage $v$.

Bond Graph Modeling ■ 39

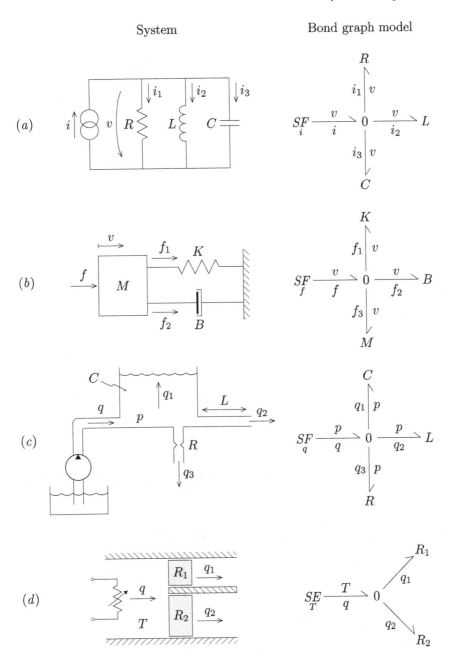

Figure 2.13 Simple examples of a 0 junction for (a) electrical, (b) mechanical, (c) hydraulic and (d) thermal systems.

**40** ■ An Introduction to Bond Graph Modeling with Applications

Figure 2.14  Modulated transformer and gyrator, $MTF$ and $MGY$.

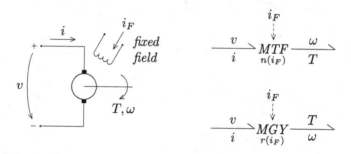

Figure 2.15  Ideal DC servomotor with separate fixed field current and bond graph representation.

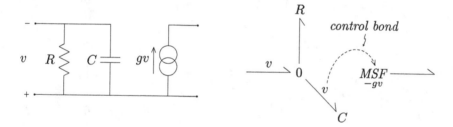

Figure 2.16  Transistor model: equivalent circuit for common emitter and bond graph representation.

## 2.4  EXAMPLES OF APPLICATION OF BOND GRAPHS

We consider here the application of the previous concepts in four examples.

### 2.4.1  Example 1: Electrical Circuit

Let us consider the electrical circuit represented in Figure 2.17 and the corresponding bond graph in Figure 2.18.

We note in Figure 2.17 that the voltage source $v$ and the resistance $R_1$ and inductance $L_1$ share the same current $i_1$. Therefore, they are in series, or using the bond graph formalism, they are connected through a 1 junction. This corresponds to the path from to ① to ②. The same

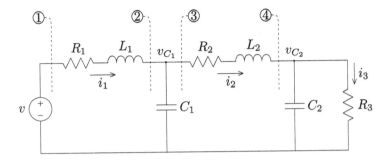

Figure 2.17 Analysis of a electrical circuit in order to derive the bond graph model.

applies to the resistance $R_2$ and inductance $L_2$ since they share the same current $i_2$. Therefore, they are connected through a 1 junction describing the path from ③ to ④. These two structures involve a parallel connection by the capacitance $C_1$ in between. Therefore, we have a 0 junction between ② and ③. Finally, after ④ we note the parallel connection of $C_2$ and $R_3$ that is modeled by a second 0 junction. The overall description with the bond graph formalism is depicted in Figure 2.18.

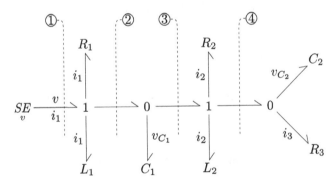

Figure 2.18 The bond graph of the electrical system in Figure 2.17.

### 2.4.2 Example 2: Translational Mechanical System

We have the translational mechanical system represented in Figure 2.19. The corresponding bond graphs are illustrated in Figures 2.20 and 2.21, for the analogies 'Option 1' and 'Option 2', respectively.

If we adopt 'Option 1', then the inertias $M_1$ and $M_2$ are connected by 0 junctions since we have the same velocity in both sides of each

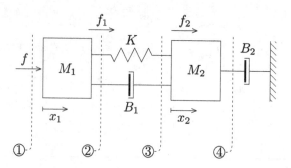

Figure 2.19  Analysis of the translational mechanical system.

element. The two cases correspond to the paths from ① to ② and from ③ to ④, respectively. Moreover, the elements $K$ and $B_1$ are in parallel and are connected also by a 0 junction. However, the path from ② to ③ is described by a 1 junction since they share the same flow, that is, forces $f_1$ and $f_2$ are identical. If we consider that the energy source is a $SF$ delivering a driving force $f$, then we can connect it to the 0 junction at ①.

If we adopt 'Option 2' and compare with the previous case, then 1 junctions are converted to 0 junctions and *vice versa*. Moreover, The $SF$ is converted to a $SE$.

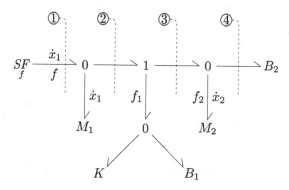

Figure 2.20  The bond graph of the mechanical system in Figure 2.19 using 'Option 1' analogy.

### 2.4.3  Example 3: Hydraulic System

This example consists of the hydraulic system represented in Figure 2.22. The corresponding bond graph is represented in Figure 2.23.

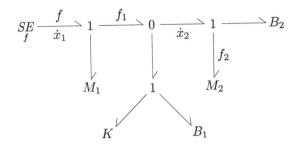

**Figure 2.21** The bond graph of the system in Figure 2.19 using the 'Option 2' analogy.

We have the hydraulic inertance $L_1$ and resistance $R_1$ in series since they share the same flow $q_1$. Identically, the inertance $L_2$ and resistance $R_2$ are in series since they share the flow $q_2$. In these cases the bond graph representations corresponds to a 1 junction and to the paths from ① to ② and from ③ to ④, respectively. The hydraulic capacitance $C$ between ② and ③ is modeled by a 0 junction, since both terminals have the same pressure $p_2$. If we consider that the hydraulic pump is a $SF$ delivering the flow $q_1$, then it connects to the first 1 junction at ①.

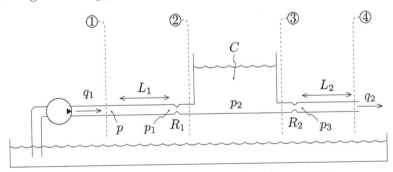

**Figure 2.22** Analysis of a hydraulic system.

### 2.4.4 Example 4: Thermal System

In this case we consider the thermal system represented in Figure 2.24 and the corresponding bond graph in Figure 2.25.

We have a thermal energy source $SF$ delivering the flow $q$ connecting at ①. We note clearly two parallel paths for flows $q_1$ and $q_2$ at ①, respectively, that can be represented by a 0 junction. For the upper path the thermal resistance $R_1$ and capacitance $C_1$ are in series since

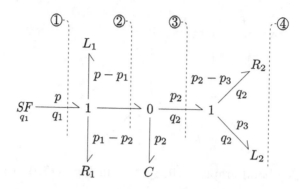

Figure 2.23 The bond graph of the hydraulic system of Figure 2.22.

they share the same flow $q_1$ and, therefore, they are connected by a 1 junction. For the lower path we note at ④ that $q_2$ passing through $R_2$ flows partially to $C_2$ and the rest to $R_3$. Therefore, this corresponds to a parallel connection represented by a 0 junction. Finally, the flow $q_3$ passes through $R_3$ and $C_3$ and this corresponds to a 1 junction.

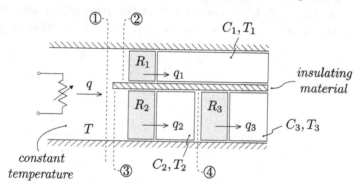

Figure 2.24 Example of a thermal system.

## 2.5 CAUSALITY

One important feature of bond graphs is the 'causality', that is represented graphically by a vertical bar (also called causal stroke) placed at only one of the ends of each bond. As we shall see, there are rules for assigning the proper causality to a given element or port, and rules for the precedence among them. The input/output causality is not usually discussed in explicit terms with other mathematical descriptions, such as with transfer functions or equation formulation. However, the causality

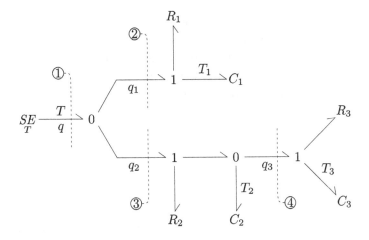

Figure 2.25 The bond graph of the thermal system of Figure 2.24.

assignment is very useful in bond graph modeling since it clarifies the input/output roles of effort and flow. In other words, the positions of the causality strokes, show which of the power variables are dependent and which are not, showing clearly some 'degeneracy' in the system.

The causal stroke convention is illustrated in Figure 2.26. In the left bond graph diagram the causal stroke indicates that (i) flow is the input to the bond, while (ii) effort is the output of the bond. In other words, in the left diagram $\mathcal{E}_1$ imposes flow and $\mathcal{E}_2$ imposes effort, while in the right diagram $\mathcal{E}_1$ imposes effort and $\mathcal{E}_2$ imposes flow.

One should distinguish between causality and power flow, since they have completely distinct meanings. The power flow arrow indicates the assumed direction of positive flow on a given bond. On the other hand, the causal stroke indicates which of the system variables (flow or effort) is assumed to be the input and output of the bond. For example, in the

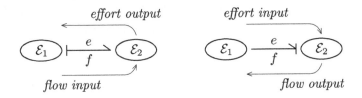

Figure 2.26 Causality between two elements where it is supposed that $\mathcal{E}_1$ is delivering power to $\mathcal{E}_2$. Left diagram $\mathcal{E}_1$ imposes flow and $\mathcal{E}_2$ imposes effort. Right diagram $\varepsilon_1$ imposes effort and $\mathcal{E}_2$ imposes flow.

diagrams in Figure 2.26, it is supposed that $\mathcal{E}_1$ is delivering power to $\mathcal{E}_2$. Indeed, the bond graph representations of Figure 2.27 consider the same causality as for those in Figure 2.26, but have the inverse power flow. Therefore, when discussing merely causality we must not consider the half-arrow by reading them simply as shown in Figure 2.28.

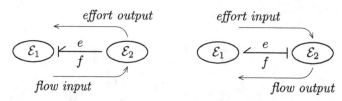

Figure 2.27 Causality between two elements where it is supposed that $\mathcal{E}_2$ is delivering power to $\mathcal{E}_1$. Left diagram $\mathcal{E}_1$ imposes flow and $\mathcal{E}_2$ imposes effort. Right diagram $\mathcal{E}_2$ imposes effort and $\mathcal{E}_2$ imposes flow.

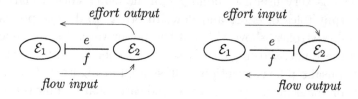

Figure 2.28 Causality between two elements without considering the power flow. Left diagram $\mathcal{E}_1$ imposes flow and $\mathcal{E}_2$ imposes effort. Right diagram $\mathcal{E}_2$ imposes effort and $\mathcal{E}_2$ imposes flow.

### 2.5.1 Causality: Energy Sources

The assignment of causality to the energy sources is defined by its type. The effort and flow sources have only the 'effort output' and the 'flow output' causalities shown in Figure 2.29. The half-arrow is just for one possible direction of the power, but can be the inverse. Therefore, we can say that:

- The effort sources have only the 'effort output' causality.

- The flow sources have only the 'flow output' causality.

**Figure 2.29** Causality of energy sources. The half-arrow is just for one possible direction of the power, but can be the inverse.

### 2.5.2 Causality: Energy Stores (Integral Causality)

In the case of the effort stores there is an 'intrinsic' causal form, namely the integral causality. This causality follows the way we consider the energy storage:

- The effort store integrates the input effort $e$ to give an effort accumulation $e_a(t) = \int_0^t e(\tau) d\tau$. Therefore, the output flow $f$ is a function of the properties of the device and $e_a$, keeping in mind that:
  - If the flow is a non-linear function of the accumulated effort, then $f = \varphi_L^{-1}(e_a)$.
  - If the flow is a linear function of the accumulated effort, then $f = L^{-1} e_a$.

- The flow store integrates the input flow $f$ to give an flow accumulation $f_a(t) = \int_0^t f(\tau) d\tau$. Therefore, the output effort $e$ is a function of the properties of the device and $f_a$, so that:
  - If the effort is a non-linear function of the accumulated flow, then $e = \varphi_C^{-1}(f_a)$.
  - If the effort is a linear function of the accumulated flow, then $e = C^{-1} f_a$.

The diagrams in Figure 2.30 show the effort (top) and flow (bottom) stores. As before, the half-arrow is just for one possible direction of the power, but can be the inverse.

To summarize, the preferred causality for:

- effort stores is 'effort input/flow output',
- flow stores is 'flow input/effort output',

as represented in Figure 2.31.

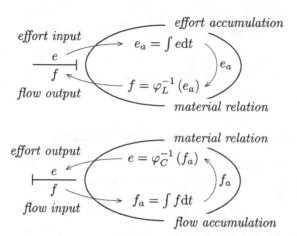

**Figure 2.30** Causality: effort (top) and flow (bottom) stores.

$$\longmapsto C \qquad \longrightarrow\!\!\mid L$$

**Figure 2.31** Causality: flow store (left) and effort store (right).

### 2.5.3 Causality: Dissipators

Dissipators have a static constitutive relation constraining the effort and flow in the device. For this reason, there is no preferred causality associated with dissipators and both causal forms are feasible as shown in Figure 2.32.

$$\longmapsto R \qquad \longrightarrow\!\!\mid R$$

**Figure 2.32** Causality: dissipators.

### 2.5.4 Causality: High-Order Elements

The memristor mentioned in sub-section 2.2 and Figures 2.1 and 2.3 can be also represented in the bond graph formalism. We can have:

- impulse or accumulated effort $(e_a)$ controlled $f_a = F(e_a)$ which implies that $\frac{df_a}{dt} = \frac{dF(e_a)}{dp} \frac{de_a}{dt} \Rightarrow f = W(e_a) e$,

- charge or accumulated flow $(f_a)$ controlled $e_a = G(f_a)$ which implies that $\frac{de_a}{dt} = \frac{dG(f_a)}{dq} \frac{df_a}{dt} \Rightarrow e = M(f_a) f$, where $F$ and $G$ are linear or non-linear functions. The bond graph representations are shown in Figures 2.33 and 2.34 for the impulse and charge controlled memristors, respectively.

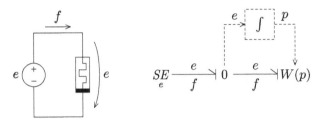

Figure 2.33  Memristor: impulse controlled.

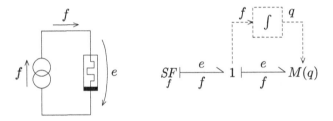

Figure 2.34  Memristor: charge controlled.

### 2.5.5  Causality: Junctions

The causality associated with a junction is determined by noting that all bonds share a common:

- flow on a effort junction. Therefore, only one bond on a effort junction can have flow as an output variable.
- effort on a flow junction. Therefore, only one bond on a flow junction can have effort as an output variable.

Figure 2.35 depicts the corresponding bond graph representations. The half-arrow power direction is arbitrary.

### 2.5.6  Causality: Two-ports

The causalities that are possible in the $TF$ and $GY$ elements follow from their definitions.

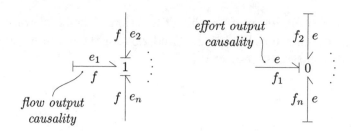

Figure 2.35 Causality for effort junction (left) and flow junction (right).

- The *TF* only modifies the ratio by which effort and flow jointly transport energy. Therefore, the causality of the output bond is the same as that of the input bond.

- The *GY* reverses the roles of effort and flow. Hence, the causality of the output bond is the opposite to that on the input bond.

Figure 2.36 depicts the bond graph representations for the *TF* and *GY*. The half-arrow power direction in the diagrams can be also the inverse.

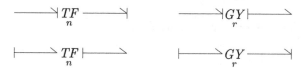

Figure 2.36 Causality of two-port elements: *TF* (left) and *GY* (right).

## 2.6 ASSIGNING CAUSALITY TO A BOND GRAPH

The causality layout of a bond graph is obtained by having in mind that we have a kind of hierarchy in the rules. For one-port elements we have:

1. The source elements must obey the prescribed causalities. However, we must note that in the mechanical systems often there is some difficulty in distinguishing between *SE* and *SF*. Therefore, in these cases a careful approach is needed.

2. The storage elements have the preferred integral causality.

3. The dissipators can have any causality.

In this line of thought, the causality of a bond graph may be obtained as follows:

- Assign the causality to the energy source elements.

- Assign the preferred (integral) causality to the energy storage elements.

- Follow the consequences of the previous steps by assigning causality that the sources force upon the other system elements by means of the system two ports and junctions. This step can be adopted either immediately after each of the previous steps, or only after completing both.

- Search for any remaining dissipators that have not yet some causal assignments. In this case, assign them an arbitrary causality for completing the procedure.

After applying this procedure, we can have three cases, depending upon the topology of the system. Therefore, the steps discussed previously can:

- Be sufficient and completely specify the causality of the bond graph elements.

- Be insufficient to determine the causality of the bond graph. Therefore, we must adopt the fourth step to the remaining dissipators, that is, we need to assign arbitrary causality in order to complete the causality assignment in the bond graph.

- Lead to a 'causal conflict' that occurs when the system model does not comply with the application of the causality rules. For example, this problem emerges when not all storage devices can be given integral causality. This is due to the presence of compatibility constraints involving only:

    - flow stores and effort sources
    - effort stores and flow sources

Any of these cases implies that there is linear dependence among the stored energy variables. Loosely speaking, it means that the adopted model overlooks some extra dynamic effects that can eventually be neglected in other system structures, but not in the present case. This problem can often be surpassed by including one or more additional elements in some 'proper place' reflecting more precisely the system properties. However, determining both the element and its location requires some experience and has to be carried carefully. In practical terms, even if the extra element(s) are included the problem of 'causal conflict' means that some anomalous dynamical behavior, with high transient values for the effort or flow may occur.

### 2.6.1 Example of Causality Assignment: Electrical Circuits

Let us consider the electrical circuit discussed in sub-section 2.4.1 with the layout of Figure 2.17. The bond graph representation was given in Figure 2.18. If we apply the causality rules we obtain the bond graph represented in Figure 2.37. In this case all rules of causality were compatible with each other.

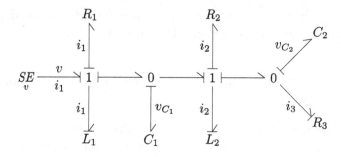

Figure 2.37  Assigning causality to the electrical circuit of Figure 2.17.

As mentioned before, we can have in some cases either a 'indeterminate', or a 'causal conflict' assignment. The first occurs when after using all rules we have not yet completed the causality assignment for all bond graph. This case poses no problem and we just assign some arbitrary causality to one of the remaining elements and then proceed. The second occurs when the assignment of the rules is not compatible due to the circuit topology. This case means that the diagram has to include additional elements representing more assertively the physical behavior of the real system.

Let us now consider the simple electrical circuit and bond graph depicted in Figure 2.38. In this case we verify a causality 'indeterminate' assignment. We just consider any causality either to $R_1$ or to $R_2$ and then we proceed without further problems.

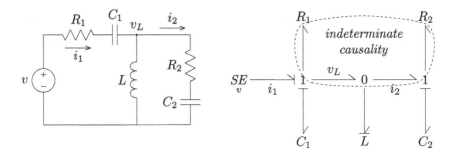

**Figure 2.38** An 'indeterminate' causality assignment in a simple electrical circuit.

### 2.6.2 Example of Causality Assignment: Translational Mechanical Systems

In this example we have the translational mechanical system analyzed in sub-section 2.4.2 with the layout and bond graph representations of Figures 2.19 and 2.20, respectively. After applying the causality rules we obtain the representation of Figure 2.39, where all rules are compatible.

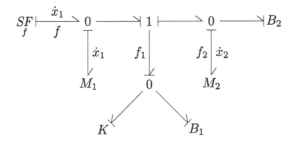

**Figure 2.39** Assigning causality to the translational mechanical system of Figure 2.19.

Let us now consider that the input action is exerted through a spring $K_0$ before inertia $M_1$ as depicted Figure 2.40. This spring can be seen as a model of the non-rigid surface of the inertia $M_1$. The corresponding bond graph shows a 'causal conflict'. We must note that the order of application of rules often dictates the point where we detect the 'causal conflict'. However, that does not means that there is a particular problem there, since a different order of application of the rules will eventually lead to the 'causal conflict' elsewhere. Therefore, users have to analyze the model searching for the point where the model oversimplifies the real system. In our case, if we consider that the non-rigid surface of inertia $M_1$ is modeled by a spring $K_0$ and a friction $B_0$, then we have the representation in Figure 2.41. We observe that in this case there is no 'causal conflict'.

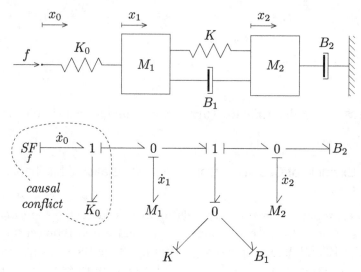

Figure 2.40 A 'causal conflict' in a translational mechanical system. The surface of inertia $M_1$ is modeled by means of $K_0$.

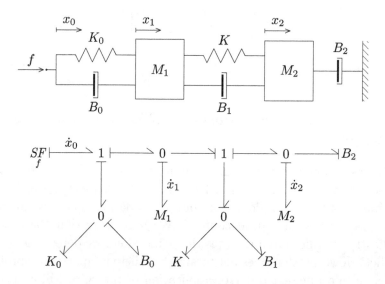

Figure 2.41 A translational mechanical system without 'causal conflict'. The surface of inertia $M_1$ is modeled by means of $K_0$ and $B_0$.

### 2.6.3 Example of Causality Assignment: Hydraulic System

In this example we have the hydraulic system analyzed in sub-section 2.4.3 with the layout and bond graph representations of Figures 2.22 and 2.23, respectively. We observe a 'causal conflict' after applying the causality rules (Figure 2.42). This problem can be solved if, for example, we consider that the hydraulic pump has some internal leakage modeled by a resistance $R_0$ in parallel. This gives the bond graph of Figure 2.43.

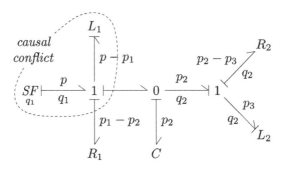

Figure 2.42 A 'causal conflict' in the hydraulic system of Figure 2.22.

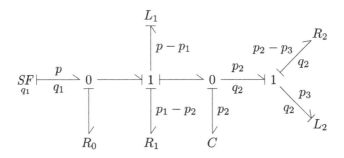

Figure 2.43 The hydraulic system of Figure 2.22 without 'causal conflict'. The pump is considered to have an internal leakage $R_0$.

### 2.6.4 Example of Causality Assignment: Thermal System

In this example we have the thermal system analyzed in sub-section 2.4.4 with the layout and bond graph representations of Figures 2.24 and 2.25, respectively. In this example there is no 'causal conflict', as shown in Figure 2.44.

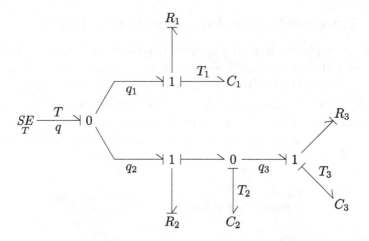

Figure 2.44 Assigning causality to the thermal system of Figure 2.24.

CHAPTER 3

# Electrical Systems

## 3.1 INTRODUCTION

The bond graph modeling technique was introduced in Chapter 2. Detailed analysis of electrical circuits can be found in subsections 2.4.1 and 2.6.1.

In the bond graph representation one interconnects elements by means of two variables, denoted as effort $e$ and flow $f$, so that the power is given by $P = e \cdot f$. The assignment of a given variable as effort or flow is essentially abstract. Nonetheless, in electrical systems the usual is to consider voltage and current for effort and flow variables, respectively.

For constructing the bond graph, we can consider the following steps.

1. Find the system elements that have the same voltage and connect them by means of a 0 junction,

2. Find the system elements that share the same current and connect them by means of a 1 junction,

3. Connect the 0 and 1 junctions to form a single bond graph,

4. Simplify the bond graph by eliminating any unnecessary element or junction,

5. Assign causality to the bond graph.

The sources of effort are the system imposed voltages. In the same line of thought, the sources of flow are the system imposed currents. A 1 junction (or a 0 junction) is associated with a given current (or a voltage). The dissipator, flow and effort store elements consist of the

**58** ■ An Introduction to Bond Graph Modeling with Applications

TABLE 3.1 Bond graph elements for electrical systems.

| Element | | Bond Graph |
|---|---|---|
| Effort | voltage, $v$ | |
| Flow | current, $i$ | |
| Effort source | voltage source, $v$ | $SE \xrightarrow{v}$ |
| Flow source | current source, $i$ | $SF \xmapsto{i}$ |
| Effort store (inductor) | $i = \frac{1}{L}\int_0^t v(\tau)d\tau$ or $i = \frac{1}{L}\lambda$ | $\xrightarrow{} L$ |
| Flow store (capacitor) | $v = \frac{1}{C}\int_0^t i(\tau)d\tau$ or $v = \frac{1}{C}q$ | $\xmapsto{} C$ |
| Dissipator (resistor) | $v = Ri$ | $\xrightarrow{} R$ $\xmapsto{} R$ |

resistance, capacitance and inductance. Table 3.1 summarizes the bond graph elements for electrical systems.

For example, let us consider the electrical system of Figure 3.1, where $R_1$, $R_2$ and $R_3$ denote resistances, $C$ represents capacitance, and $L$ corresponds to the inductance. Moreover, $v_1$, $v_2$ and $v_3$ stand for voltages and $i_1$, $i_2$ and $i_3$ are currents. The bond graph representation shown in Figure 3.2 includes one $SE$, two 1 junctions and one 0 junction. Also, the bond graph accommodates all causality rules.

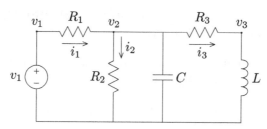

Figure 3.1 Example of an electrical circuit.

## 3.2 SOLVED PROBLEMS

**Problem 3.1.** Consider the electrical circuit shown in Figure 3.3, where $R_1$ and $R_2$ are resistances, $C_1$ and $C_2$ stand for the capacitors and $L$ represents the inductance. Draw the corresponding bond graph and discuss the causality.

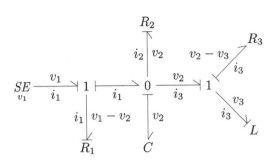

Figure 3.2 Bond graph representation the electrical circuit of Figure 3.1.

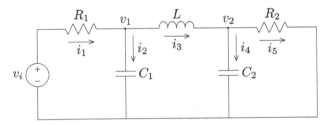

Figure 3.3 The electrical circuit of Problem 3.1.

**Resolution:** Figure 3.4 illustrates the construction of the bond graph for the proposed electrical circuit. We consider three steps for obtaining the diagram as represented by the lines ①, ② and ③. The fixed voltage source $v_i$ and the resistance $R_1$, together with the electrical connection crossing ①, share the same current $i_1$. Therefore, they are connected by an 1 junction.

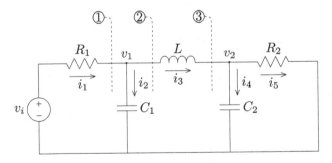

Figure 3.4 Analysis of the electrical system of Problem 3.1.

The capacitor $C_1$ is placed in parallel with the electrical connections crossing ① and ②. This structure imposes a flow constraint to the system and, therefore, they are interconnected by a 0 junction resulting in the partial bond graph in Figure 3.5.

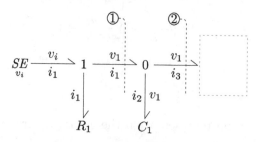

Figure 3.5 The partial bond graph of Problem 3.1.

The inductance $L$ shares a common current ($i_3$) with the conductive path crossing ② and ③ being, therefore, connected by an 1 junction. The remaining two components, resistance $R_1$ and capacitor $C_2$, and the electrical connection crossing ③ share a common voltage ($v_2$) and different currents ($i_3 = i_4 + i_5$). We can connect them through a 0 junction. The complete bond graph is presented in Figure 3.6, where the lines ①, ② and ③ clarify the relationship with the circuit of Figure 3.4.

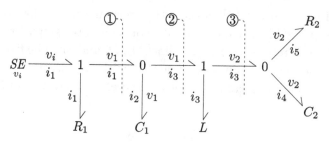

Figure 3.6 The bond graph of Problem 3.1.

The procedure of assigning the causality to the bond graph starts by applying the required causality to the source $v_i$ and the preferred causality to the storage devices ($C_1$, $L$ and $C_2$). This step results in the bond graph of Figure 3.7.

The causality not yet assigned to the remaining components can be found via the required causality of the system multi-ports. In this case, the causal strokes of the capacitors $C_1$ and $C_2$ that are connected to the 0 junctions and the inductance $L$ connected to the 1 junction give us the base to assign the remaining causality. After completing this procedure we obtain the bond graph of Figure 3.8.

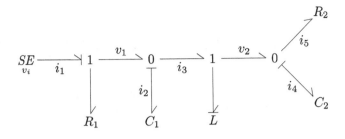

Figure 3.7 The bond graph with causality assigned to the source and stores (causality not yet assigned to the rest of the elements and junctions).

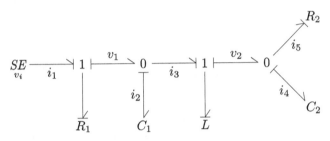

Figure 3.8 The complete bond graph of Problem 3.1.

If we adopt the standard model we obtain the set of equations (with zero initial conditions):

$$\begin{cases} v_i = R_1 i_1 + v_1 \\ v_1 = \frac{1}{C_1} \int_0^t i_2(\tau)\, d\tau \\ i_1 = i_2 + i_3 \\ i_3 = \frac{1}{L} \int_0^t (v_1(\tau) - v_2(\tau))\, d\tau \\ v_2 = \frac{1}{C_2} \int_0^t i_4(\tau)\, d\tau \\ i_3 = i_4 + i_5 \\ v_2 = R_2 i_5 \end{cases}$$

The standard block diagram is represented in Figure 3.9. We verify that the bond graph modeling leads to a much more rational and conservative representation.

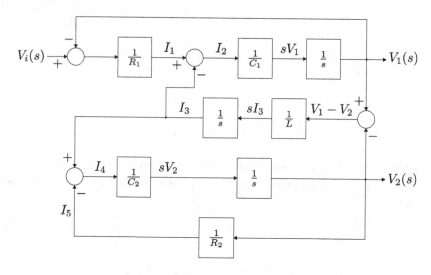

Figure 3.9  The block diagram of Problem 3.1 in the Laplace domain.

**Problem 3.2.** Consider the electrical circuit in Figure 3.10 that uses an ideal electrical transformer. Moreover, let $R_1$ and $R_2$ denote resistances, $L_1$ and $L_2$ represent the inductances and $C$ stand for the capacitor. Find a bond graph representation of this circuit and discuss the causality.

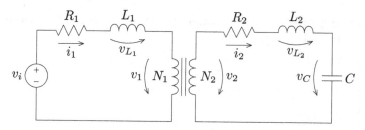

Figure 3.10  The electrical circuit of Problem 3.2.

**Resolution:** In order to construct the bond graph of the circuit in Figure 3.10 the most important task is to analyze how the elements are interconnected. Recall that bond graph elements are connected by means of multi-port junctions and two-port elements.

We can observe that the current $i_1$ is the common flow variable to the passive elements $R_1$ and $L_1$, the fixed voltage source $v_i$ and the primary winding of the electrical transformer. Therefore, they are in series and are connected by a 1 junction. The secondary winding of the electrical transformer is in series with the elements $R_2$, $L_2$ and $C$ since they share the same current $i_2$. As before, these elements are connected through a 1 junction.

The electrical transformer scales the power variables since $\frac{v_1}{v_2} = \frac{i_2}{i_1} = \frac{N_1}{N_2}$, where $N_1$ and $N_2$ represent the number of turns in the transformer windings. Therefore, the secondary voltage $v_2$ is proportional to the primary voltage $v_1$ and the primary current $i_1$ is proportional to the secondary current $i_2$, using the same proportionality constant $\frac{N_2}{N_1}$. For that reason this electrical component is modeled by the energy conservative two-port bond graph element $TF$. The result is the bond graph shown in Figure 3.11.

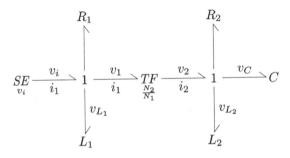

Figure 3.11 The bond graph of Problem 3.2.

The classical alternative dynamic equations of Problem 3.2 (considering zero initial conditions):
$$\begin{cases} v_i = R_1 i_1 + v_{L_1} + v_1 \\ i_1 = \frac{1}{L_1} \int_0^t v_{L_1}(\tau)\, d\tau \\ i_2 = \frac{1}{L_2} \int_0^t v_{L_2}(\tau)\, d\tau \\ v_C = \frac{1}{C} \int_0^t i_2(\tau)\, d\tau \\ v_2 = R_2 i_2 + v_{L_2} + v_C \\ \frac{v_1}{v_2} = \frac{i_2}{i_1} = \frac{N_1}{N_2} \end{cases}$$

The standard block diagram derived for the circuit is shown in Figure 3.12. The bond graph for the same system leads to a much more clear and natural representation.

The procedure for assigning the causality to the bond graph of Figure 3.11 is as follows:

1. Start by assigning the required causality to the effort source $v_i$.

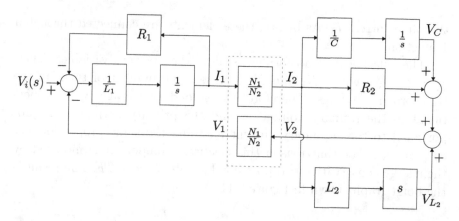

Figure 3.12  The block diagram of Problem 3.2 in the Laplace domain.

2. The second step is to assign the preferred causality to the storage devices, in this case the elements $L_1$, $L_2$ and $C$. The result of these first two steps is shown in Figure 3.13.

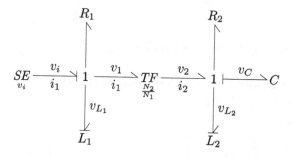

Figure 3.13  The bond graph with causality assigned to the source and stores (causality not yet assigned to the rest of the elements and junctions).

3. The next step can be applied to either 1 junctions since both inductances $L_1$ and $L_2$ fulfill the respective multi-port causal constraints. This enables the assignment of the causality to the remaining bonds of the junction. For the 1 junction located on the right side of the bond graph, we can assign the causal strokes to the bonds that connect to $R_2$ and the output port of $TF$.

4. For the element $TF$, if the causality in one port is known, then the causality of the other port is automatically established. Hence,

the causality can be assign between the input port of $TF$ and the 1 junction.

5. The causality not yet assigned (element $R_1$) can be found as described in the third step.

This sequence results in the bond graph shown in Figure 3.14 that has causal conflicts. We can verify that both elements $L_1$ and $TF$ have a flow output causality when connected to the (left) 1 junction. This conflict can be addressed by modifying the physical model.

Figure 3.15 shows the electrical circuit with a leakage resistance $R_L$ connected (in parallel) to the primary winding of the transformer. The correspondent bond graph is depicted in Figure 3.16 where the addition of the 0 junction solves the causality conflict. The same outcome could also be achieved using a leakage capacitor instead of the resistance $R_L$, or by adding both in parallel with the primary winding of the transformer.

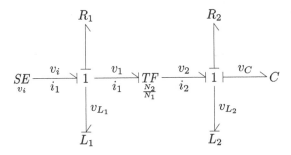

Figure 3.14 The bond graph of Problem 3.2 including the causality.

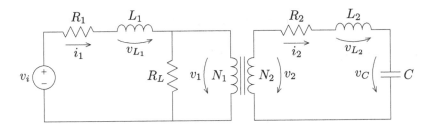

Figure 3.15 The modified electrical circuit of Problem 3.2.

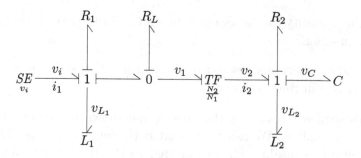

**Figure 3.16** The bond graph of the circuit in Figure 3.16.

## 3.3 PROPOSED EXERCISES

**Exercise 3.1.** Consider the electrical circuit shown in Figure 3.17(a), where $R_i$ $(i = 1, 2)$ are resistances and $C_i$ stand for the capacitors. The corresponding bond graph in Figure 3.17(b), where the symbols $\alpha$, $\beta$ and $\gamma$ represent electrical elements of the circuit.

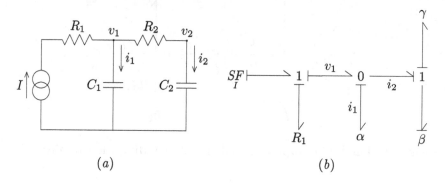

**Figure 3.17** The (a) electrical circuit and (b) bond graph of Exercise 3.1.

The correct response is one of the following cases:

A) $\alpha = R_2$, $\beta = C_1$, $\gamma = C_2$

B) $\alpha = C_1$, $\beta = R_2$, $\gamma = C_2$

C) $\alpha = C_2$, $\beta = R_2$, $\gamma = C_1$

D) $\alpha = C_2$, $\beta = C_1$, $\gamma = R_2$

**Exercise 3.2.** Consider the electrical circuit of Figure 3.18 and the corresponding bond graph represented in Figure 3.19. The circuit consists of the voltage source $v_i$, the inductances $L_1$ and $L_2$, the capacitors $C_1$ and $C_2$, and the resistances $R_1$, $R_2$ and $R_3$. The symbols $\alpha$, $\beta$, $\gamma$, $\delta$ and $\varepsilon$ denote components of the bond graph. The symbols $\alpha$ and $\delta$ are:

**A)** $\alpha = R_2$, $\delta = R_1$

**B)** $\alpha = R_1$, $\delta = R_2$

**C)** $\alpha = SE$, $\delta = R_1$

**D)** $\alpha = C_1$, $\delta = R_1$

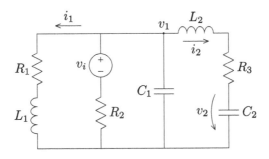

Figure 3.18 The electrical circuit of Exercise 3.2.

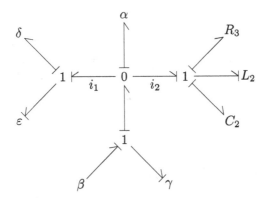

Figure 3.19 The bond graph of Exercise 3.2.

**Exercise 3.3.** Consider the electrical circuit shown in Figure 3.20(a) and the corresponding bond graph sketched in Figure 3.20(b). Let $R_i$ ($i = 1, 2, 3$), $L_j$ ($j = 1, 2$) and $C$ denote the resistances, the inductances and the capacitor, respectively. The symbols $\alpha$, $\beta$, $\gamma$ and $\delta$ denote elements of the bond graph. Then we can say that the symbols:

**A)** $\alpha$ and $\gamma$ represent 0 junctions

**B)** $\beta$ represents a 1 junction and $\gamma$ a 0 junction

**C)** $\alpha$ and $\gamma$ represent 1 junctions

**D)** $\beta$ represents a 0 junction and $\gamma$ a 1 junction

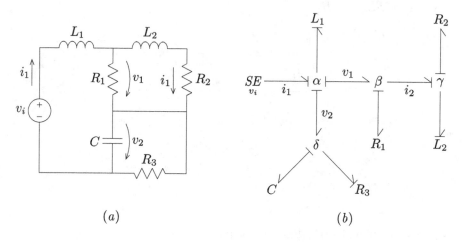

Figure 3.20 The (a) electrical circuit and (b) bond graph of Exercise 3.3.

**Exercise 3.4.** Figure 3.21 shows a electrical circuit where the symbols $R_i$ ($i = 1, 2$), $L_i$ and $C_j$ ($j = 1, 2, 3$) stand for resistances, capacitors and inductances, respectively. Additionally, consider the corresponding bond graph illustrated in Figure 3.22 where the symbols $\alpha$, $\beta$, $\gamma$, $\delta$ and $\varepsilon$ denote junctions.

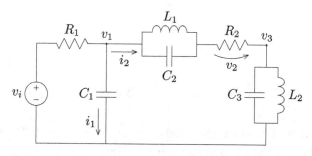

Figure 3.21 The electrical circuit of Exercise 3.4.

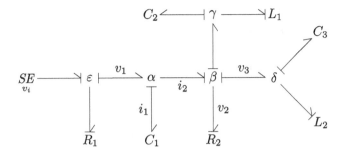

Figure 3.22 The bond graph of Exercise 3.4.

In this case the symbols:

**A)** $\beta$ and $\delta$ represent 1 junctions

**B)** $\beta$ and $\delta$ represent 0 junctions

**C)** $\beta$ represents a 1 junction and $\delta$ a 0 junction

**D)** $\beta$ represents a 0 junction and $\delta$ a 1 junction

**Exercise 3.5.** Consider the electrical circuit shown in Figure 3.23 that uses an ideal electrical transformer with the transformation ratio $n = \frac{N_1}{N_2}$. Moreover, let $R_1$, $R_2$ and $R_3$ represent resistances, $C_1$ and $C_2$ stand for capacitors and $L$ be a inductance. Figure 3.24 shows the corresponding bond graph where the symbols $\alpha$, $\beta$, $\gamma$ and $\delta$ denote elements of the system. For the symbol $\alpha$ we can say:

**A)** $\alpha = C_1$

**B)** $\alpha = L$

**C)** $\alpha = R_2$

**D)** $\alpha = C_2$

**Exercise 3.6.** Consider the bond graph in Figure 3.25, where $SE$ stands for effort source, $R$ denotes dissipator elements, $C$ and $L$ represent flow and effort stores, respectively, and $TF$ is a two-port transformer element with the transformation ratio $\frac{1}{n}$.

1. Assign the causality of the bond graph.

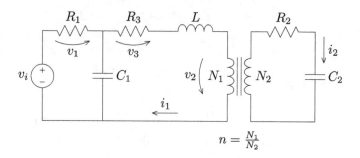

Figure 3.23 The electrical circuit of Exercise 3.5.

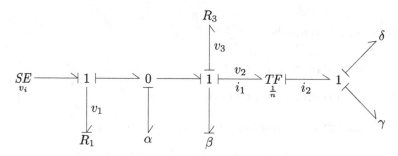

Figure 3.24 The bond graph of Exercise 3.5.

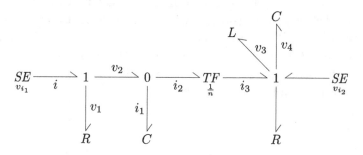

Figure 3.25 The bond graph of Exercise 3.6.

2. Figure 3.26 shows an electrical circuit derived from the bond graph of Exercise 3.6. Complete the circuit with all the effort and flow variables represented on the bond graph.

**Exercise 3.7.** Draw a schematic diagram of an electrical circuit that can be modeled by the bond graph given in Figure 3.27. Consider that $SF$ is the current source, $R_1$, $R_2$ and $R_3$ are the resistances, the symbols $C_1$ and $C_2$ represent the capacitors and $L$ denotes the inductance.

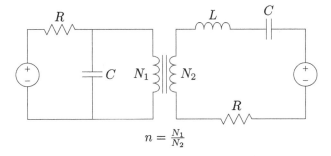

Figure 3.26 The electrical circuit of Figure 3.25.

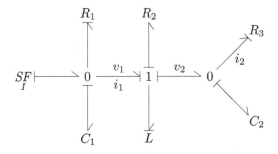

Figure 3.27 The bond graph of Exercise 3.7.

**Exercise 3.8.** Consider the electrical circuit shown in Figure 3.28 that includes two resistances ($R_1$ and $R_2$), two capacitors ($C_1$ and $C_2$) and a fixed voltage source $v_i$. Draw the corresponding bond graph taking into account the causality of the components.

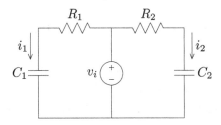

Figure 3.28 The electrical circuit of Exercise 3.8.

**Exercise 3.9.** Consider the electrical circuit illustrated in Figure 3.29, where the symbols $R_i$ ($i = 1, 2$), $C_i$ and $L_i$ denote the resistances, the capacitors and the inductances, respectively. Draw the bond graph including all electrical variables, and discuss the causality.

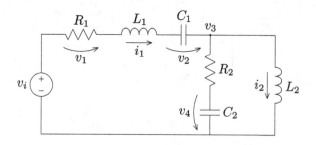

Figure 3.29 The electrical circuit of Exercise 3.9.

**Exercise 3.10.** Consider the electrical circuit shown in Figure 3.30, where $R_i$ ($i = 1, 2, 3, 4, 5$) are resistances, $L_1$ and $L_2$ stand for inductances, $C_1$ and $C_2$ represent capacitors and $v_i$ is the applied voltage. Draw the bond graph of the circuit and assign the corresponding causality.

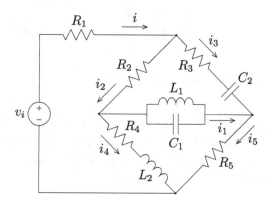

Figure 3.30 The electrical circuit of Exercise 3.10.

**Exercise 3.11.** Figure 3.31 shows a electrical circuit that includes a voltage source $v_i$, a current source $I$ and an ideal electrical transformer with the transformation ratio $n = \frac{N_1}{N_2}$. Moreover, consider that $R_1$ and $R_2$ represent resistances, $C_1$, $C_2$ and $C_3$ stand for the capacitors and $L$ is the inductance. Obtain the bond graph of the system and discuss the causality.

**Exercise 3.12.** Consider the electrical circuit shown in Figure 3.32, where the dissipator elements are denoted by $R_i$ and $R_j$ ($j = 1, 2, 3$), the flow stores by $C_k$ ($k = 1, 2$) and the effort stores by $L_k$. Furthermore, the output of the flow source ($i_2$) follows the direct relation $i_2 = Kv_1$,

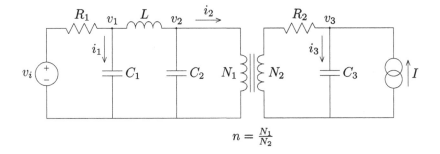

Figure 3.31 The electrical circuit of Exercise 3.11.

where $v_1$ is the voltage across the capacitor $C_1$ and $K$ is a constant gain. Find the corresponding bond graph and assign the causality.

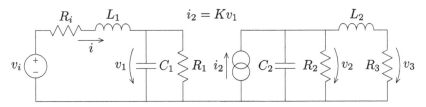

Figure 3.32 The electrical circuit of Exercise 3.12.

CHAPTER 4

# Mechanical Systems

## 4.1 INTRODUCTION

Chapter 2 introduced the bond graph graphical modeling tool. We can find examples of mechanical systems in subsections 2.4.2 and 2.6.2.

This technique interconnect elements by means of two variables, denoted as effort $e$ and flow $f$, so that their product gives power $P = e \cdot f$. The assignment of a given variable as effort or flow is purely abstract and, therefore, we can have different choices. In mechanical systems two distinct options have been adopted by authors and researchers. Some works consider linear/angular velocity and force/torque for effort and flow variables, respectively, while others consider the inverse case. The two alternatives are valid and we shall mention them as options 1 and 2. While in this book we consider mainly the 'Option 1', we include some problems solved also with 'Option 2'.

For constructing the bond graph, we can consider the following steps.

- Option 1, {linear/angular velocity, force/torque}→{effort, flow}

    1. Find the system elements that move with the same velocity and connect them by means of a 0 junction,
    2. Find the system elements that have the same applied force/torque and connect them by means of a 1 junction,
    3. Connect the 0 and 1 junctions to form a single bond graph,
    4. Simplify the bond graph by eliminating any unnecessary element or junction,
    5. Assign causality to the bond graph.

- Option 2, {force/torque, linear/angular velocity}→{effort, flow}

  1. Find the system elements that move with the same velocity and connect them by means of a 1 junction,
  2. Find the system elements that have the same applied force/torque and connect them by means of a 0 junction,
  3. Connect the 0 and 1 junctions to form a single bond graph,
  4. Simplify the bond graph by eliminating any unnecessary element or junction,
  5. Assign causality to the bond graph.

With 'Option 1' the sources of effort are the system imposed velocities, and the sources of flow are the system imposed forces/torques. A 1 junction (or a 0 junction) is associated with a given force/torque (or a linear/angular velocity). Therefore, the dissipator, flow and effort store elements consist of the damper, mass/inertia and spring. Table 4.1 summarizes the bond graph elements for mechanical systems when adopting Option 1: {linear/angular velocity, force/torque}→{effort, flow}.

**TABLE 4.1** {linear/angular velocity, force/torque}→{effort, flow}.

| Element | Linear | Rotational | Bond Graph |
|---|---|---|---|
| Effort | velocity, $\dot{x}$ | velocity, $\dot{\theta} = \omega$ | |
| Flow | force, $f$ | torque, $T$ | |
| Effort source | linear velocity source, $\dot{x}$ | angular velocity source, $\omega$ | $SE \longrightarrow\!\!\!\!\!\!\!\!\dot{x}$  $SE \longrightarrow\!\!\!\!\!\!\!\!\omega$ |
| Flow source | force source, $f$ | torque source, $T$ | $SF \vdash\!\!\!\longrightarrow\!\!\!\!\!\!\!\!f$  $SF \vdash\!\!\!\longrightarrow\!\!\!\!\!\!\!\!T$ |
| Effort store (compliance) | $f = K \int_0^t \dot{x}(\tau)\,\mathrm{d}\tau$ or $f = Kx$ | $T = K \int_0^t \omega(\tau)\,\mathrm{d}\tau$ or $T = K\theta$ | $\longrightarrow\!\!\!\mid K$ |
| Flow store (inertia) | $\dot{x} = \frac{1}{M}\int_0^t f(\tau)\,\mathrm{d}\tau$ or $\dot{x} = \frac{1}{M}p$ | $\omega = \frac{1}{M}\int_0^t T(\tau)\,\mathrm{d}\tau$ or $\omega = \frac{1}{M}h$ | $\vdash\!\!\!\longrightarrow M$ |
| Dissipator (friction) | $f = B\dot{x}$ | $T = B\omega$ | $\vdash\!\!\!\longrightarrow B$  $\longrightarrow\!\!\!\mid B$ |

With 'Option 2' the sources of effort are the system imposed forces/torques, and the sources of flow are the imposed velocities. Moreover, a 1 junction (or a 0 junction) is associated with a given linear/angular velocity (or a force/torque). Therefore, the dissipator, flow and effort store elements consist of the damper, spring and mass/inertia. Table 4.2 summarizes the bond graph elements for mechanical systems when adopting Option 2: {force/torque, linear/angular velocity} →{effort, flow}.

**TABLE 4.2**  {force/torque, linear/angular velocity}→{effort, flow}.

| Element | Linear | Rotational | Bond Graph |
|---|---|---|---|
| Effort | force, $f$ | torque, $T$ | |
| Flow | velocity, $\dot{x}$ | velocity, $\dot{\theta} = \omega$ | |
| Effort source | force source, $f$ | torque source, $T$ | $SE \xrightarrow{f}$ $SE \xrightarrow{T}$ |
| Flow source | linear velocity source, $\dot{x}$ | angular velocity source, $\omega$ | $SF \xrightarrow{\dot{x}}$ $SF \xrightarrow{\omega}$ |
| Effort store (inertia) | $\dot{x} = \frac{1}{M}\int_0^t f(\tau)\,d\tau$ or $\dot{x} = \frac{1}{M}p$ | $\omega = \frac{1}{M}\int_0^t T(\tau)\,d\tau$ or $\omega = \frac{1}{M}h$ | $\longrightarrow M$ |
| Flow store (compliance) | $f = K\int_0^t \dot{x}(\tau)\,d\tau$ or $f = Kx$ | $T = K\int_0^t \omega(\tau)\,d\tau$ or $T = K\theta$ | $\longmapsto K$ |
| Dissipator (friction) | $f = B\dot{x}$ | $T = B\omega$ | $\longmapsto B$ $\longrightarrow B$ |

Let us consider the mechanical system of Figure 4.1, where $M_1$ and $M_2$ denote masses, $K_1$ and $K_2$ represent the springs stiffness constants, $B_1$ and $B_2$ correspond to the dampers viscous friction coefficients, and $\dot{x}_1$, $\dot{x}_2$ and $\dot{x}_3$ stand for linear velocities.

Following the convention of Table 4.1 (i.e., Option 1: {linear/angular velocity, force/torque}→{effort, flow}) we obtain the bond graph in Figure 4.2. On the other hand, if we adopt the convention of Table 4.2 (i.e., Option 2: {force/torque, linear/angular velocity}→{effort, flow}), then we obtain the bond graph in Figure 4.3.

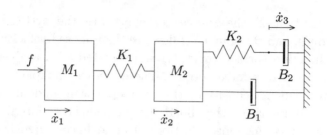

Figure 4.1 Example of a mechanical system.

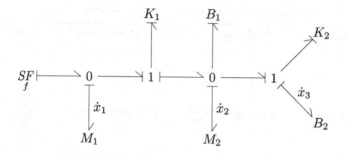

Figure 4.2 The bond graph of the mechanical system of Figure 4.1, when adopting Option 1: {linear/angular velocity, force/torque}→{effort, flow}.

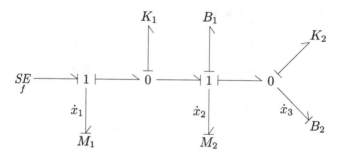

Figure 4.3 The bond graph of the mechanical system of Figure 4.1, when adopting Option 2: {force/torque, linear/angular velocity}→{effort, flow}.

## 4.2  SOLVED PROBLEMS

**Problem 4.1.** The mechanical system of Figure 4.4 comprises the masses $M_1$ and $M_2$, the dampers with viscous friction coefficients $B_1$, $B_2$ and $B_3$, and the springs with stiffness constants $K_1$, $K_2$ and $K_3$. The applied force is represented by $f$ and the linear displacements are denoted by $x_1$, $x_2$ and $x_3$. Draw the bond graph and include the causality of the components.

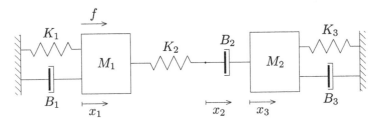

Figure 4.4   The linear mechanical system of Problem 4.1.

**Resolution:** Consider the system represented in Figure 4.5. The applied force $f$, the mass $M_1$, the elements $K_1$ and $B_1$, and the link crossing the dashed line ① share the same velocity $\dot{x}_1$. Therefore, they are connected by means of a 0 junction (Figure 4.6(a)). The mass $M_2$, the elements $K_3$ and $B_3$, and the link crossing the dashed line ② are another set of elements that share the same velocity, in this case $\dot{x}_3$. Figure 4.6(b) shows the 0 junction that connects these elements.

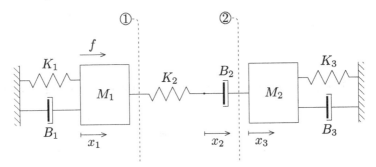

Figure 4.5   Analysis of the mechanical system of Problem 4.1.

The spring $K_2$, the damper $B_2$, the link crossing ① and the link crossing ② share a common force, but have different velocities. To express the relationships between these elements a 1 junction is used, as shown in the bond graph of Figure 4.7. Attaching the three partial bond graphs using the common links that cross the lines ① and ②, leads to the diagram in Figure 4.8.

The application of the required causality to the flow source $f$ and the preferred causality to the storage devices ($M_1$, $M_2$, $K_1$, $K_2$ and $K_3$)

**80** ■ An Introduction to Bond Graph Modeling with Applications

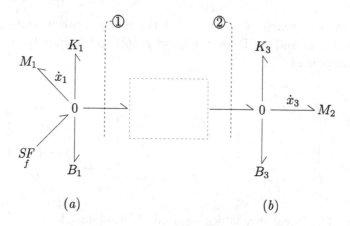

Figure 4.6 The partial bond graphs of Problem 4.1 for elements that share the same (a) velocity $\dot{x}_1$ and (b) velocity $\dot{x}_3$.

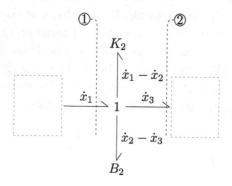

Figure 4.7 The partial bond graph of Problem 4.1 for elements with distinct velocities.

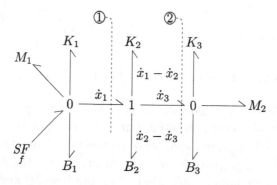

Figure 4.8 The bond graph of Problem 4.1 without the causality assigned.

results in the bond graph depicted in Figure 4.9. The causality of the stores fulfill the requirements to assign the causality to the remaining bonds via the causality rules of multi-port elements. In more detail, the masses $M_1$ and $M_2$ have effort as the output variable and are connected to a 0 junction. The spring with stiffness constant $K_2$ is connected to a 1 junction and has flow as the output variable. The complete bond graph is shown in Figure 4.10.

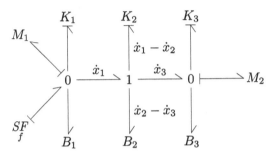

Figure 4.9 The bond graph with causality assigned to the source and stores (causality not yet assigned to the rest of the elements and junctions).

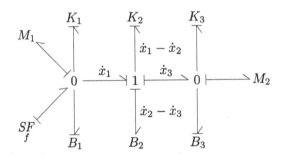

Figure 4.10 The complete bond graph of Problem 4.1.

If we adopt the standard model we obtain the set of differential equations:
$$\begin{cases} f = M_1\ddot{x}_1 + B_1\dot{x}_1 + K_1 x_1 + K_2(x_1 - x_2) \\ K_2(x_1 - x_2) = B_2(\dot{x}_2 - \dot{x}_3) \\ 0 = M_2\ddot{x}_3 + B_3\dot{x}_3 + K_3 x_3 + B_2(\dot{x}_3 - \dot{x}_2) \end{cases}$$

The standard block diagram is represented in Figure 4.11. We verify that the bond graph modeling leads to a much more rational and conservative representation.

**82** ■ An Introduction to Bond Graph Modeling with Applications

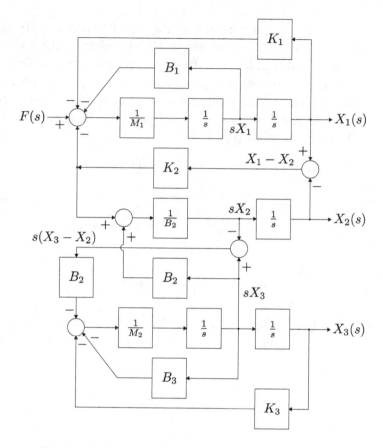

Figure 4.11  The block diagram of Problem 4.1 in the Laplace domain.

**Problem 4.2.** Consider the rotational mechanical system shown in Figure 4.12, where $T_m$ is the applied torque, $\theta_m$, $\theta_T$ and $\theta_L$ represent angular displacements, $K$ is the spring stiffness constant (Hooke's law), $B_m$, $B$ and $B_L$ denote the rotational viscous friction coefficients and, $J_m$ and $J_L$ are inertias. The system has a gear, with $N_i$ ($i = 1, 2$) number of teeth, that provide the transformation ratio $n = \frac{N_1}{N_2}$. Draw the bond graph of the system considering the causality of the components.

**Resolution:** The procedure to derive the bond graph model of Problem 4.2 can start by grouping all the elements that share the same effort variable (i.e., the angular velocities). The applied torque $Tm$, the inertia $J_m$ and the element $B_m$ share the same angular velocity $\dot{\theta}_m$. Therefore, they are connected by means of a 0 junction (Figure 4.13(a)). Figure 4.13(b) shows the 0 junction that connects the elements $K$ and $B$

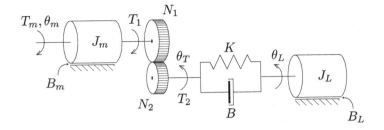

Figure 4.12 The rotational mechanical system of Problem 4.2.

that share the same relative angular velocity $\dot{\theta}_T - \dot{\theta}_L$. The last set of elements that share the same effort ($\dot{\theta}_L$) is composed by the inertia $J_L$ and the element $B_L$ and is shown in Figure 4.13(c). All the bond graphs in Figure 4.13 have a bond that models the link connecting that set of elements to the rest of the system.

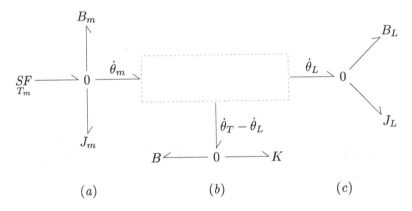

Figure 4.13 The partial bond graphs of Problem 4.2 for elements that share the same angular velocities: (a) $\dot{\theta}_m$, (b) $\dot{\theta}_T - \dot{\theta}_L$ and (c) $\dot{\theta}_L$.

The next step is to connect the partial bond graphs using 1 junctions to model the elements that share the same flow variable, but have different effort variables. Consider the system representation shown in Figure 4.14. The elements $K$ and $B$, and the links crossing the dashed lines ② and ③ share a common torque that leads to the bond graph in Figure 4.15.

The final bond graph (Figure 4.16) is obtained by connecting the links that cross the lines ① and ② by means of a two-port element $TF$ that models the gear as a mechanical transformer, since $\dot{\theta}_T = n\dot{\theta}_m$ and $T_1 = nT_2$.

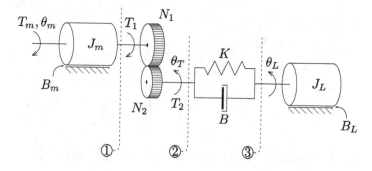

Figure 4.14  Analysis of the mechanical system of Problem 4.2.

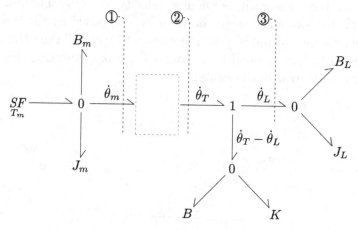

Figure 4.15  The partial bond graph of Problem 4.2 for elements with distinct angular velocities.

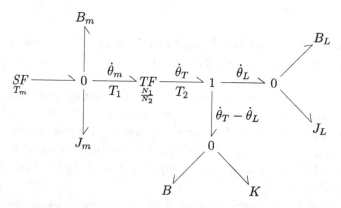

Figure 4.16  The bond graph of Problem 4.2 without the causality.

The procedure for assigning the causality to the bond graph of Figure 4.16 is as follows:

1. Assign the required causality to the flow source $T_m$.

2. Follow with the assignment of the preferred causality to the storage devices ($J_m$, $B$ and $J_L$).

3. For each junction with an element with required or preferred causality, that meets the multi-port causal constraints, propagate the causality assignment through the graph. The inertias $J_m$ and $J_L$ fulfill this criteria.

These steps result in the bond graph shown in Figure 4.17 that has no causal conflict.

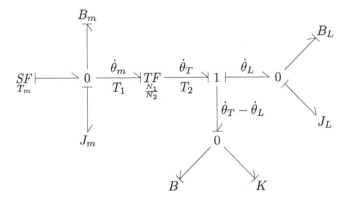

Figure 4.17 The bond graph of Problem 4.2 including the causality.

If we adopt the standard model we obtain the set of differential equations:

$$\begin{cases} T_m = J_m \ddot{\theta}_m + B_m \dot{\theta}_m + T_1 \\ T_2 = K(\theta_T - \theta_L) + B\left(\dot{\theta}_T - \dot{\theta}_L\right) \\ T_2 = J_L \ddot{\theta}_L + B_L \dot{\theta}_L \\ \frac{T_1}{T_2} = \frac{\theta_T}{\theta_m} = \frac{N_1}{N_2} \end{cases}$$

Figure 4.18 shows the bond graph for Problem 4.2 if we adopt Option 2, i.e., the correspondence of effort and flow to torque and angular velocity, respectively.

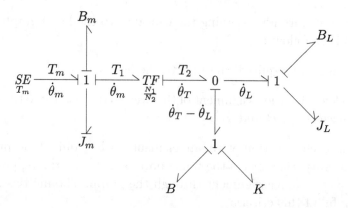

Figure 4.18  The Option 2 bond graph of Problem 4.2.

## 4.3  PROPOSED EXERCISES

**Exercise 4.1.** Consider the linear mechanical system and the corresponding bond graph shown in Figure 4.19. The system is driven by force $f$ applied to mass $M$, causing the displacement $x$. The springs stiffness constants are represented by $K_1$ and $K_2$, and the dampers viscous friction coefficients by $B_1$ and $B_2$. The symbols $\alpha$, $\beta$, $\delta$, $\mu$ and $\varepsilon$ represent one-port elements of the bond graph. We can say that:

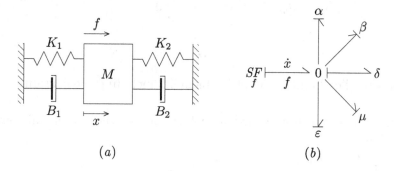

(a)            (b)

Figure 4.19  The (a) linear mechanical system and (b) bond graph of Exercise 4.1.

A) $\alpha = B_1$, $\beta = B_2$, $\delta = M$, $\mu = K_2$, $\varepsilon = K_1$

B) $\alpha = M$, $\beta = K_1$, $\delta = K_2$, $\mu = B_1$, $\varepsilon = B_2$

C) $\alpha = K_2$, $\beta = K_1$, $\delta = B_2$, $\mu = B_1$, $\varepsilon = M$

D) $\alpha = B_1$, $\beta = K_2$, $\delta = K_1$, $\mu = M$, $\varepsilon = B_2$

**Exercise 4.2.** Consider the linear mechanical system shown in Figure 4.20(a), where $f$ is the applied force, $x_1$ and $x_2$ denote horizontal displacements, $K_1$ and $K_2$ stand for spring stiffness constants (Hooke's law), $B$ is the viscous friction coefficient and $M_1$ and $M_2$ are masses. The symbols $\alpha$, $\beta$ and $\gamma$ included in the bond graph of Figure 4.20(b) correspond to:

**A)** $\alpha = K_1,\ \beta = M_1,\ \gamma = M_2$

**B)** $\alpha = M_1,\ \beta = K_1,\ \gamma = M_2$

**C)** $\alpha = K_1,\ \beta = M_2,\ \gamma = M_1$

**D)** $\alpha = M_1,\ \beta = M_2,\ \gamma = K_1$

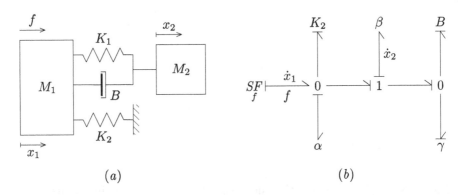

**Figure 4.20** The (a) mechanical system and (b) bond graph of Exercise 4.2.

**Exercise 4.3.** Consider the mechanical system of Figure 4.21 and the bond graph in Figure 4.22. Let $f$ be the applied force, $x_1$, $x_2$ and $x_3$ the horizontal displacements, $K_1$ and $K_2$ the springs stiffness constants (Hooke's law), $B_1$ and $B_2$ the viscous friction coefficients and $M_1$, $M_2$ and $M_3$ the masses. In this case, it follows that:

**A)** $\alpha = B_2,\ \beta = B_1,\ \delta = K_1,\ \varepsilon = K_2$

**B)** $\alpha = B_1,\ \beta = B_2,\ \delta = K_2,\ \varepsilon = K_1$

**C)** $\alpha = K_2,\ \beta = K_1,\ \delta = B_2,\ \varepsilon = B_1$

**D)** $\alpha = K_1,\ \beta = K_2,\ \delta = B_2,\ \varepsilon = B_1$

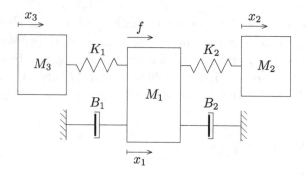

Figure 4.21 The linear mechanical system of Exercise 4.3.

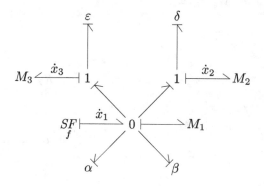

Figure 4.22 The bond graph of Exercise 4.3.

**Exercise 4.4.** Consider the mechanical system shown in Figure 4.23, where $f$ is the applied force, $x_1$, $x_2$ and $x_3$ represent horizontal displacements, $K_1$, $K_2$ and $K_3$ are the spring stiffness constants (Hooke's law), $B_1$, $B_2$ and $B_3$ denote the viscous friction coefficients and $M_1$, $M_2$ and $M_3$ are masses.

1. For the bond graph of Figure 4.24 it comes that:

   **A)** $\alpha = M_1$, $\beta = B_1$, $\gamma = K_1$, $\delta = M_2$, $\mu = K_2$, $\varepsilon = M_3$, $\chi = K_3$
   **B)** $\alpha = K_1$, $\beta = B_1$, $\gamma = M_1$, $\delta = K_3$, $\mu = M_3$, $\varepsilon = K_2$, $\chi = M_2$
   **C)** $\alpha = K_1$, $\beta = B_1$, $\gamma = M_1$, $\delta = K_2$, $\mu = B_2$, $\varepsilon = K_3$, $\chi = B_3$
   **D)** $\alpha = K_1$, $\beta = B_1$, $\gamma = M_1$, $\delta = K_2$, $\mu = M_2$, $\varepsilon = K_3$, $\chi = M_3$

2. Find an analogous electrical circuit of the mechanical system using the force $\rightarrow$ current analogy.

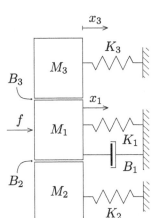

Figure 4.23 The linear mechanical system of Exercise 4.4.

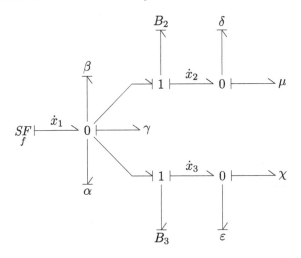

Figure 4.24 The bond graph of Exercise 4.4.

**Exercise 4.5.** Consider the mechanical system shown in Figure 4.25, where $f$ is the applied force, $x_i$ ($i = 1, 2, 3$) represents the displacements, $K_i$ ($i = 1, 2$) denote the springs stiffness constants and $M_i$ ($i = 1, 2, 3$) are the rigid bodies masses. The bond graph of the system is also sketched in Figure 4.26, where the symbols $\alpha$, $\beta$, $\gamma$, $\delta$ and $\varepsilon$ denote elements of the system. In the bond graph we can say:

1.  A) $\delta = M_2, \varepsilon = M_3$
    B) $\gamma = M_1, \varepsilon = M_2$
    C) $\alpha = M_3, \varepsilon = M_2$
    D) $\gamma = M_2, \varepsilon = M_3$

2.  A) $\beta = B_2$
    B) $\beta = B_1$
    C) $\beta = M_2$
    D) $\beta = K_2$

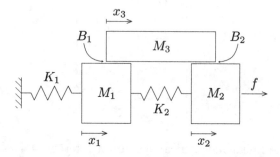

Figure 4.25 The linear mechanical system of Exercise 4.5.

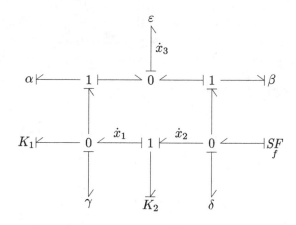

Figure 4.26 The bond graph of Exercise 4.5.

**Exercise 4.6.** Consider the mechanical system shown in Figure 4.27, where forces $f_1$ and $f_2$ are applied to rigid bars with neglectable mass. Let $x_1$, $x_2$ and $x_3$ denote linear displacements, $K_1$ and $K_2$ represent the springs stiffness constants, $B_1$ and $B_2$ stand for the viscous friction coefficients and $M$ is the mass. Figure 4.28 depicts the corresponding bond graph.

1. The symbols $\alpha$, $\beta$ and $\gamma$ represent which type of junction?

2. Assign the causality to the bond graph.

3. Determine the mathematical model and the block diagram representation of the mechanical system.

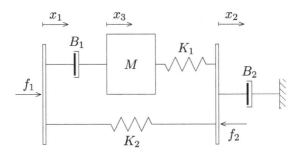

Figure 4.27  The linear mechanical system of Exercise 4.6.

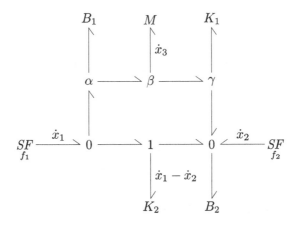

Figure 4.28  The bond graph of Exercise 4.6.

**Exercise 4.7.** Figure 4.29 shows a rotational mechanical system that includes springs with the stiffness constants $K_1$ and $K_2$, dampers with the viscous friction coefficients $B_1$ and $B_2$, and a inertia denoted by $J$. Let $T$, $\theta_1$ and $\theta_2$ be the externally applied torque and the angular displacements, respectively. The corresponding bond graph is shown in Figure 4.30, where the symbols $\alpha$, $\beta$, $\gamma$, $\delta$ and $\varepsilon$ represent elements of the mechanical system. For the symbol $\gamma$ we can write:

**A)** $\gamma = B_1$

**B)** $\gamma = K_1$

**C)** $\gamma = K_2$

**D)** $\gamma = J$

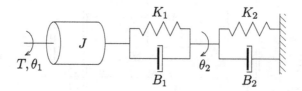

**Figure 4.29** The rotational mechanical system of Exercise 4.7.

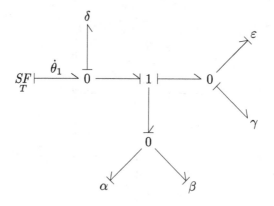

**Figure 4.30** The bond graph of Exercise 4.7.

**Exercise 4.8.** Consider the mechanical system shown in Figure 4.31, where $T$ is the externally applied torque, $\theta_1$, $\theta_2$ and $\theta_3$ represent angular displacements, $K_1$ and $K_2$ are the springs stiffness constants (Hooke's law), $J_1$ and $J_2$ are inertias and, $B_i$ ($i = 1, 2, 3, 4$) denote the rotational viscous friction coefficients.

1. Complete the partial bond graph in Figure 4.32.
2. Assign the causality to the bond graph of the exercise.

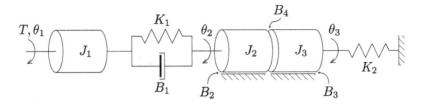

Figure 4.31 The rotational mechanical system of Exercise 4.8.

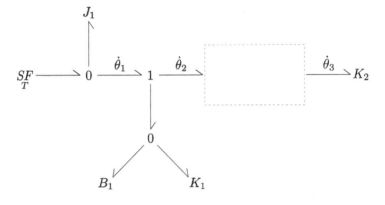

Figure 4.32 The partial bond graph of Exercise 4.8.

**Exercise 4.9.** Consider the mechanical system represented in Figure 4.33, where $T_m$ and $T_L$ are the torques applied by an actuator and requested by the load, respectively. The system has a gear with transformation ratio $n = \frac{N_1}{N_2}$, where $N_i$ ($i = 1, 2$) is the number of teeth of the $i$-th gear-wheel. It can be concluded that the symbol $\alpha$ in the bond graph of Figure 4.34 corresponds to:

**A)** $\alpha = J_1$

**B)** $\alpha = B_1$

**C)** $\alpha = J_2$

**D)** $\alpha = B_2$

**Exercise 4.10.** Consider the mechanical system shown in Figure 4.35, where $T$ is the input torque, $B_i$ ($i = 1, 2, 3$) denote the rotational viscous friction coefficients, $J_i$ are inertias and $K_j$ ($j = 1, 2$) stand for the springs stiffness constants (Hooke's law). Moreover, the angular displacements

**94** ■ An Introduction to Bond Graph Modeling with Applications

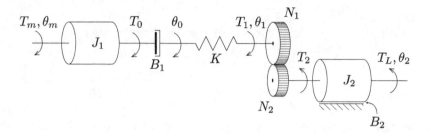

Figure 4.33 The rotational mechanical system of Exercise 4.9.

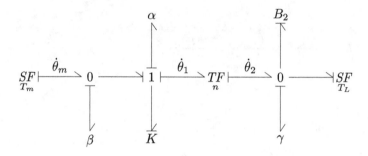

Figure 4.34 The bond graph of Exercise 4.9.

are represented by the symbol $\theta_k$ ($k = 0, 1, 2, 3$). The system has two belt drives with transformation ratios $\frac{r_{11}}{r_{12}}$ and $\frac{r_{21}}{r_{22}}$, where the constants $r_{jj}$ are the pulleys radii. Draw the bond graph and assign the causality.

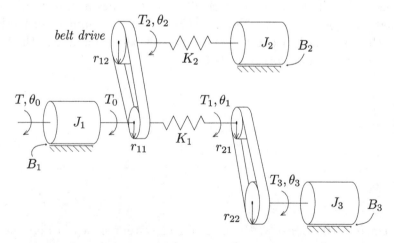

Figure 4.35 The rotational mechanical system of Exercise 4.10.

**Exercise 4.11.** Consider the bond graph of a rotational mechanical system shown in Figure 4.36 that has a causality conflict on the right side 1 junction. The bond graph follows the Option 2 convention where {torque, angular velocity}→{effort, flow}. Let the symbols $SE$, $B$, $K$ and $J$ denote the effort source, the dissipator elements, the effort and the flow stores, respectively. Moreover, the symbol $TF$ is a two-port transformer element with the transformation ratio $\frac{r_2}{r_1}$, where $r_1$ and $r_2$ represent radii. Draw the schematic of a mechanical system that can be modeled by the bond graph and having solved the causal conflict.

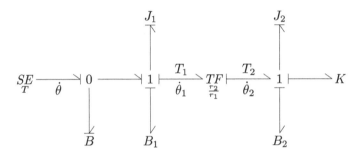

Figure 4.36  The bond graph of Exercise 4.11.

**Exercise 4.12.** Figure 4.37 shows a rotational mechanical system where $T_m$ is the torque applied by a motor and $T_L$ is the torque requested by the load. The system has a set of gears with $N_i$ ($i = 1, 2, 3$) number of teeth, respectively. Additionally, we have the following transformation ratios $n_{12} = \frac{N_1}{N_2}$ and $n_{23} = \frac{N_2}{N_3}$. Draw the bond graph of the system considering the causality of the components.

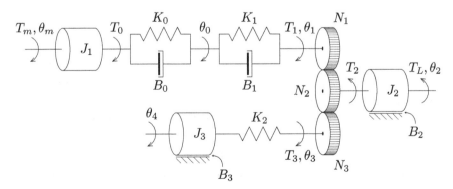

Figure 4.37  The rotational mechanical system of Exercise 4.12.

**Exercise 4.13.** Consider the mechanical system shown in Figure 4.38(a), where $T$ is the applied torque that causes the linear force $f$ and the horizontal displacement $x$. Torque $T$ is transmitted to mass $M$ by a massless pulley, without any friction or slip. Moreover, are also represented the stiffness of the springs $K_1$ and $K_2$, the viscous friction coefficient $B$ and the pulley radius $r$. The corresponding bond graph is shown in Figure 4.38(b), where the symbols $\alpha$, $\beta$, $\gamma$ and $\delta$ denote elements of the mechanical system. It follows that for the symbol $\alpha$ we can write:

A) $\alpha = B_1$

B) $\alpha = M$

C) $\alpha = K_1 + K_2$

D) $\alpha = K_2$

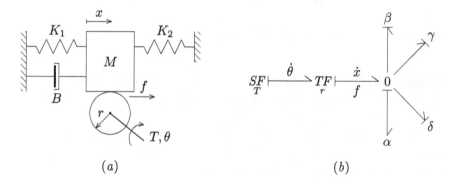

**Figure 4.38** The (a) mechanical system and (b) bond graph of Exercise 4.13.

**Exercise 4.14.** Consider the mechanical system shown in Figure 4.39, where $f$ is the applied force, $x$, $\theta_1$ and $\theta_1$ are the linear and angular displacements, $K_1$ and $K_2$ denote the springs stiffness constants, $B_1$ and $B_2$ represent the viscous friction coefficients and $M$, $J_1$ and $J_2$ stand for the linear and rotational inertias. The pulleys are frictionless and have the radius $r_1$ and $r_2$. Figure 4.40 presents the corresponding bond graph, where the symbols $\alpha$, $\beta$, $\gamma$, $\delta$ and $\varepsilon$ denote system elements.

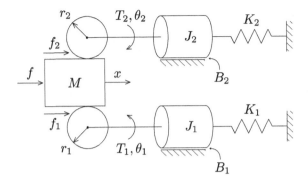

Figure 4.39 The mechanical system of Exercise 4.14.

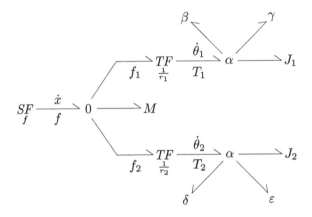

Figure 4.40 The bond graph of Exercise 4.14.

1. In the bond graph the symbol $\alpha$ represents:

    A) A multi-port element 0
    B) A multi-port element 1
    C) A two-port element $GY$
    D) Other element

2. Assign causality strokes to the bond graph.

3. In the bond graph:

    A) We observe a causal conflict
    B) We do not verify the existence of any causal conflict

**Exercise 4.15.** The mechanical system depicted in Figure 4.41 comprises a lever with arms with length $l_1$ and $l_2$ leading to the transformation ratio $n = \frac{l_1}{l_1+l_2}$. Additionally, $K_1$ and $K_2$ represent the stiffness constants of the springs, $B_1$ and $B_2$ the viscous friction coefficients and $M$ the mass. The variables $f$, $f_1$ and $f_2$, denote forces, and the variables $x$ and $x_i$ ($i = 1, 2, 3$), stand for linear displacements. Let $f$ and $x$ be the force and displacement at the input port and that the system remains linear for small angular displacements of the lever.

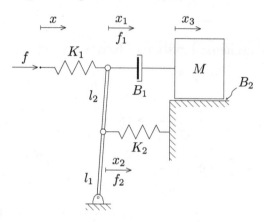

Figure 4.41  The mechanical system of Exercise 4.15.

1. In the bond graph (Figure 4.42), the symbols $\alpha$, $\beta$, $\gamma$, $\delta$ and $\varepsilon$ denote elements of the mechanical system. Then, we can say:

   (a)  **A)** $\beta = 1$, $\delta = B_1$  
        **B)** $\gamma = 1$, $\varepsilon = B_1$  
        **C)** $\varepsilon = M$, $\alpha = B_1$  
        **D)** $\gamma = 1$, $\beta = 0$,

   (b)  **A)** $\varepsilon = M$, $\gamma = 1$  
        **B)** $\gamma = 0$, $\alpha = B_1$  
        **C)** $\alpha = K_2$, $\gamma = 0$  
        **D)** $\gamma = 0$, $\beta = 0$,

2. Assign the causality to the bond graph of the exercise.

3. Regarding the causality of the bond graph, it can be said that:

   A) We verify a causal conflict  
   B) We do not have any causal conflict  
   C) Nothing can be concluded about the causality

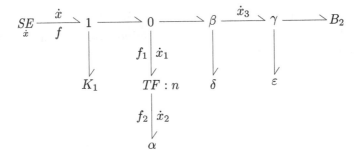

Figure 4.42 The bond graph of Exercise 4.15.

**Exercise 4.16.** Consider the mechanical system shown in Figure 4.43. The actuator torque $T$ drives the fixed pulley that is connected to the lever using a cable with elasticity. The pulley has radius $r$ and inertia $J$, while the lever has arms with lengths $l_1$ and $l_2$. Consider that $J$ is constant, regardless of the amount of wound cable and that the system remains linear for small angular displacements of the lever. Moreover, in the figure are also shown the forces $f_i$ ($i = 1, 2, 3$), the displacements $x_i$ ($i = 1, 2, 3, 4$) and $\theta$, the stiffness of the springs $K_1$ and $K_2$, the viscous friction coefficients $B_1$ and $B_2$, and the mass $M$. Represent the bond graph of the system considering the causality of the components.

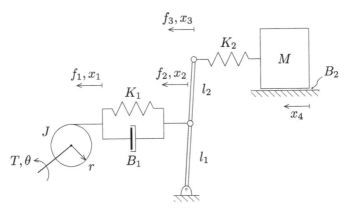

Figure 4.43 The mechanical system of Exercise 4.16.

**Exercise 4.17.** Consider the mechanical system shown in Figure 4.44, where the load, with mass $M_L$ and friction $B_L$, is connected to a worm screw by means of an association of two elements, namely a spring $K$ and damper $B$. The screw with pitch $h$ is driven by a motor with inertia

$J_m$ and rotational friction $B_m$, that provides the torque $T_m$ and the rotational displacement $\theta$. The corresponding bond graph is shown in Figure 4.45. It follows that:

**A)** $\alpha = K$

**B)** $\alpha = B$

**C)** $\alpha = M_L$

**D)** $\alpha = J_m$

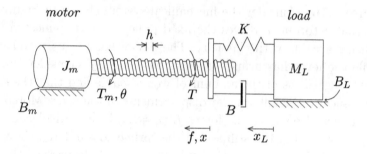

Figure 4.44 The mechanical system of Exercise 4.17.

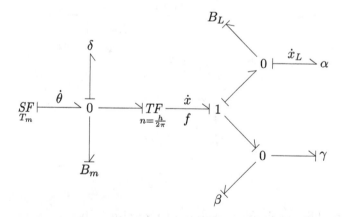

Figure 4.45 The bond graph of Exercise 4.17.

**Exercise 4.18.** Consider the mechanical system presented in Figure 4.46 involving two worm screws with pitch $h_1$ and $h_2$, respectively. Let $T_1$ and $T_2$ be the torques that drive the screws and $\theta_1$ and $\theta_2$ the related angular displacements. Furthermore, $f_1$ and $f_2$ denote forces and $x_1$ and $x_2$ stand for linear displacements. The symbol $K$ represents the spring stiffness, $B$ the linear viscous friction coefficient, $B_1$ and $B_2$ the rotational viscous friction coefficients, and $M_1$ and $M_2$ the masses. Figure 4.47 shows the bond graph that models the system, where $n_1$ and $n_2$ are conversion ratios and the symbols $\alpha$, $\beta$, $\gamma$, $\delta$, $\varepsilon$, $\zeta$ and $\eta$ denote elements of the system.

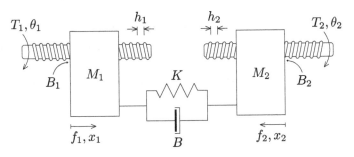

Figure 4.46  The mechanical system of Exercise 4.18.

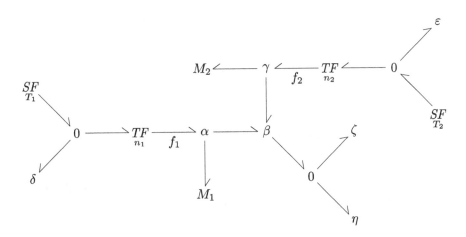

Figure 4.47  The bond graph of Exercise 4.18.

1. In the bond graph we have:

   **A)** $\zeta = B_2,\ \varepsilon = B_1$
   **B)** $\zeta = K,\ \varepsilon = B_1$
   **C)** $\zeta = K,\ \varepsilon = B_2$
   **D)** $\zeta = B_1,\ \varepsilon = B_2$

2. Assign the causality to the bond graph. Based on the result it can be stated that:

   A) There is a causal conflict in the bond graph
   B) The bond graph has no causal conflict
   C) Nothing can be said about the causality

**Exercise 4.19.** Consider the mechanical system shown in Figure 4.48. The symbols $T_i$ ($i = 1, 2$) and $\theta_i$ represent the torques and angular displacements of the gear (with $N_1$ and $N_2$ teeth) that drive the rack and pinion subsystem, where the gear has radius $r$. Additionally $f$, $x_1$ and $x_2$ represent force and linear displacements, respectively. Let $K$ be the spring stiffness constant (Hooke's law), $B$ and $B_M$ the viscous friction coefficients and $M$ the mass. Consider $n_1 = \frac{N_1}{N_2}$ and $n_2 = r$ as the transformation ratios for the gear and rank and pinion systems, respectively.

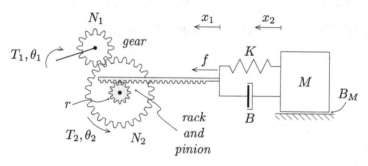

Figure 4.48  The mechanical system of Exercise 4.19.

1. We can conclude that the symbols $\alpha$, $\beta$, $\gamma$, $\delta$ and $\varepsilon$ in the bond graph of Figure 4.49 correspond to:

   **A)** $\alpha = \dot{x}_1$, $\beta = f$, $\gamma = 0$, $\delta = 1$, $\varepsilon = 0$
   **B)** $\alpha = \dot{x}_1 - \dot{x}_2$, $\beta = f$, $\gamma = 0$, $\delta = 1$, $\varepsilon = 0$
   **C)** $\alpha = f$, $\beta = \dot{x}_1$, $\gamma = 0$, $\delta = 1$, $\varepsilon = 0$
   **D)** $\alpha = \dot{x}_1$, $\beta = f$, $\gamma = 1$, $\delta = 0$, $\varepsilon = 1$

2. Assign the causality to the bond graph shown in Figure 4.49.

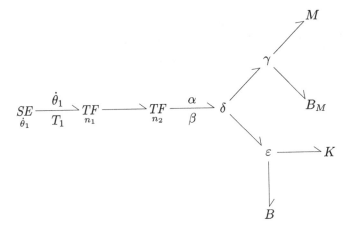

Figure 4.49 The bond graph of Exercise 4.19.

**Exercise 4.20.** The mechanical system shown in Figure 4.50 has a rack and pinion actuator with radius $r$. In the figure are also represented the applied force $f$ and the linear and angular displacements $x$, $\theta_1$ and $\theta_2$. Let $K_1$ and $K_2$ be the springs stiffness constants (Hooke's law), $B_1$ and $B_2$ the dampers viscous friction coefficients and, $M$ and $J$ the linear and rotational inertias. Sketch a bond graph that models the system considering the causality.

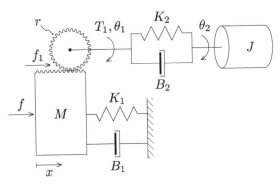

Figure 4.50 The mechanical system of Exercise 4.20.

**Exercise 4.21.** Consider the mechanical system shown in Figure 4.51 that uses a worm screw with pitch $h$, where $T_w$ and $\theta$ represent the applied torque and the angular displacement, respectively, and $B_w$ denotes the rotational viscous friction coefficient. The screw acts on the lever, with arm lengths $l_1$ and $l_2$, producing a very small angular displacement.

Consider the transformations $x_1 = n_1\theta$ and $x_2 = n_2 x_1$, where $n_1$ and $n_2$ are the transformation ratios for the worm screw and lever, respectively. Find a bond graph representation of this system and discuss the causality.

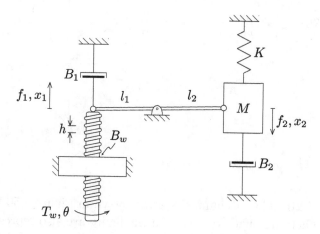

Figure 4.51  The mechanical system of Exercise 4.21.

CHAPTER 5

# Hydraulic Systems

## 5.1 INTRODUCTION

Chapter 2 introduced the bond graphs and several examples of hydraulic systems were presented in subsections 2.4.2 and 2.6.3. The usual assignment of variables as effort or flow are the pressure and flow rate, respectively. For constructing the bond graph, we can consider the following steps.

1. Find the system elements that have the same pressure and connect them by means of a 0 junction,

2. Find the system elements that share the same flow rate and connect them by means of a 1 junction,

3. Connect the 0 and 1 junctions to form a single bond graph,

4. Simplify the bond graph by eliminating any unnecessary element or junction,

5. Assign causality to the bond graph.

The sources of effort and flow are the system imposed pressures and flow rates, respectively. A 1 junction (or a 0 junction) is associated with a given flow rate (or a pressure). The dissipator, flow and effort store elements consist of the hydraulic resistance, capacitance and inertance. Table 5.1 summarizes the bond graph elements for hydraulic systems.

## 5.2 SOLVED PROBLEMS

**Problem 5.1.** Consider the hydraulic system shown in Figure 5.1, where $C_1$ and $C_2$ denote the capacitances of the tanks, $R_0$, $R_1$ and $R_2$ are the

**TABLE 5.1** Bond graph elements for hydraulic systems.

| Element | | Bond Graph |
|---|---|---|
| Effort | pressure, $p$ | |
| Flow | volume flow rate, $q$ | |
| Effort source | pressure source, $p$ | $SE \xrightharpoonup{p}$ |
| Flow source | fluid flow source, $q$ | $SF \xrightharpoonup{q}$ |
| Effort store (inertance) | $q = \frac{1}{L}\int_0^t p(\tau)d\tau$ or $q = \frac{1}{L}\Gamma$ | $\xrightharpoonup{} L$ |
| Flow store (fluid reservoir) | $p = \frac{1}{C}\int_0^t q(\tau)d\tau$ or $p = \frac{1}{C}V$ | $\xrightharpoonup{} C$ |
| Dissipator (fluid dissipator) | $p = Rq$ | $\xrightharpoonup{} R$ <br> $\xleftharpoonup{} R$ |

hydraulic resistances, $q$, $q_0$, $q_1$ and $q_2$ represent the volume flow rates, and $p_1$ and $p_2$ stand for the pressures. Draw the corresponding bond graph and discuss the causality.

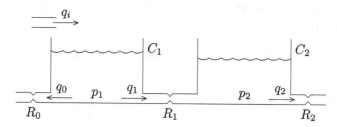

**Figure 5.1** The hydraulic system of Problem 5.1.

**Resolution:** In order to obtain the bond graph of the hydraulic system in Figure 5.1 consider the representation made in Figure 5.2, where the lines ① and ② disclose the two major steps we consider to derive the bond graph. The first step is to identify the elements connected by means of a 0 junction, i.e., with different flow rates and common pressure, that is followed by the elements with common flow rate and different pressures.

We can observe that the tank on the left has one inflow ($q_i$) and two outflows ($q_0$ and $q_1$), and that the fluid accumulation $C_1$ is the sum of these flows (law of mass conservation). While $q_i$ is provided by a constant source, the flows $q_0$ and $q_1$ result from the difference between

Hydraulic Systems ■ 107

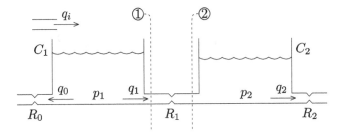

Figure 5.2 Analysis of the hydraulic system of Problem 5.1.

$p_1$ and the pressure at the other end of the hydraulic restrictions $R_0$ and $R_1$, respectively. It follows that the elements $C_1$, $R_0$, the constant flow source $q_i$ and the pipe ($q_1$) crossing the line ① are connected by a 0 junction since they all have different flow rates and the same pressure $p_1$. Note that $R_0$ is the only element on the exit pipe with $q_0$ and that it discharges to atmospheric pressure, therefore, we can connect it directly to the 0 junction. Applying the same analysis to the elements that share the common effort variable $p_2$, we can draw the bond graph of Figure 5.3.

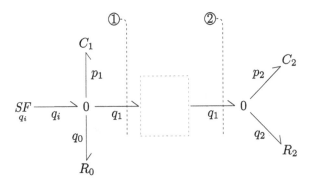

Figure 5.3 The partial bond graph of Problem 5.1.

To add the remaining element $R_1$ we must consider the different pressures on the pipes that cross the lines ① and ② ($p_1$ and $p_2$, respectively), and that the outflow $q_1$ from the tank on the left acts as an inflow to the tank on the right. Hence, the hydraulic resistance $R_1$ is connected to the two open-ended bonds of Figure 5.3 by means of a 1 junction since it has the same flow rate $q_1$ and a different pressure ($p_1 - p_2$). The result is the bond graph shown in Figure 5.4.

The procedure for assigning the causality to the bond graph of Figure 5.4 is as follows:

**108** ■ An Introduction to Bond Graph Modeling with Applications

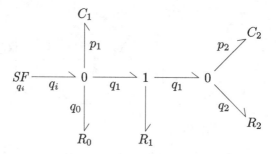

**Figure 5.4** The bond graph of Problem 5.1.

1. Start by assigning the required causality to the flow source $q_i$.

2. The second step is to assign the preferred causality to the storage devices, in this case the elements $C_1$ and $C_2$. The result of these first two steps is shown in Figure 5.5.

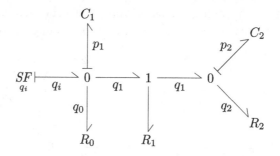

**Figure 5.5** The bond graph of Problem 5.1 after assigning the causality to source and stores.

3. The next step can be applied to either 0 junctions since we verify that both capacitances $C_1$ and $C_2$ fulfill the respective multi-port causal constraints. This enables the assignment of the causality to the remaining bonds of the two 0 junctions.

4. The causality not yet assigned (element $R_1$) is found via the 1 junction causal rules.

This sequence results in the bond graph shown in Figure 5.6 that has no causal conflict.

Hydraulic Systems ■ 109

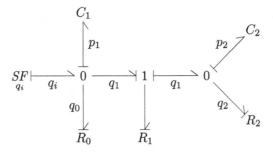

Figure 5.6 The bond graph of Problem 5.1 with the causality assigned.

If we use the standard modeling approach, we obtain the set of equations (with zero initial conditions):

$$\begin{cases} q_i = q_0 + q_1 + q_{C_1}, \; q_1 = q_2 + q_{C_2} \\ p_1 = R_0 q_0, \; p_1 - p_2 = R_1 q_1, \; p_2 = R_2 q_2 \\ p_1 = \frac{1}{C_1} \int_0^t q_{C_1}(\tau) \, d\tau \\ p_2 = \frac{1}{C_2} \int_0^t q_{C_2}(\tau) \, d\tau \end{cases}$$

The standard classical block diagram of the system is shown in Figure 5.7. We verify that the bond graph modeling results in a much more innate and restrained representation.

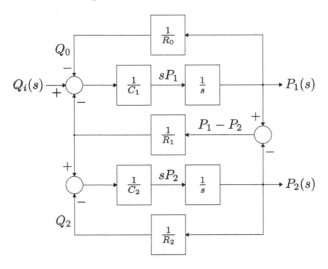

Figure 5.7 The block diagram of Problem 5.1 in the Laplace domain.

**Problem 5.2.** Consider the hydraulic system shown in Figure 5.8. The symbols $C_i$ ($i = 1, 2$), $R_j$ ($j = 1, 2, 3$) and $L_j$, represent the capacitances of the tanks, the hydraulic resistances and the inertances, respectively. Moreover, the symbol $q_j$ stands for the volume flow rates and $p_j$ are the pressures. Find the bond graph that models this system considering the causality.

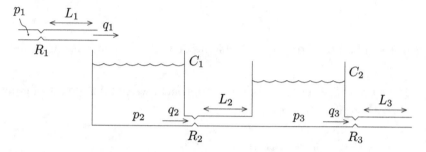

Figure 5.8 The hydraulic system of Problem 5.2.

**Resolution:** Figure 5.9 illustrates the analysis taken into account to draw the bond graph for the proposed hydraulic system. The line ① sets the bound between elements that share the same flow rate $q_1$ and elements under the same pressure $p_2$. The opposite bound of this last set of elements is delimited by the line ②, that also marks the beginning of the elements that share the same rate $q_2$. Lines ③ and ④ serve the same purpose for the flow variable $q_2$, the effort variable $p_3$ and the flow variable $q_3$, respectively.

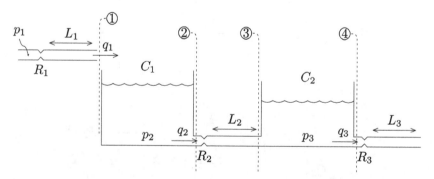

Figure 5.9 Analysis of the hydraulic system of Problem 5.2.

To start the bond graph, let us first consider elements of the system that have different pressures and a common flow rate. The pipe comprising the hydraulic resistance $R_1$ and the inertance $L_1$ has a single flow rate ($q_1$) and three different pressures: the constant input pressure $p_1$, the pressure at the discharge point of $R_1$ and the atmospheric pressure at the $L_1$ endpoint. This type of power variables constraints leads to a 1 junction connecting the mentioned elements. Moreover, the flow rate $q_1$ acts as a modulated (i.e., non constant) flow source for the tank on the left since it depends on other system variables (Figure 5.10(a)).

The lines ② and ③ enclose another section of the system traversed by a single flow rate ($q_2$) that results, therefore, on a 1 junction that connects the elements $R_2$, $L_2$ and the pipes that lead to each of the tanks. Figure 5.10(b) shows the resulting partial bond graph. The 1 junction that connects the last set of elements that share a common flow variable (i.e., the flow rate $q_3$) is shown alongside in Figure 5.10(c).

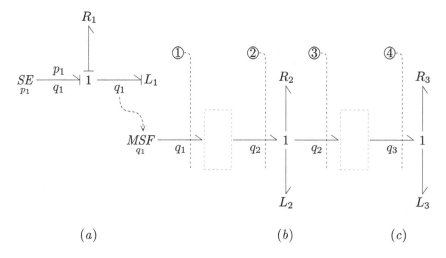

Figure 5.10 The partial bond graphs of Problem 5.2 for elements that share the same flow rate (a) $q_1$, (b) $q_2$ and (c) $q_3$.

For the final stage consider elements that have different flow rates and a common pressure. The tank on the left has the inflow $q_1$, the outflow $q_2$ and the flow accumulated in $C_1$ ($q_1 - q_2$). These power conducting bonds all share the same pressure $p_2$ and are, therefore, connected by a 0 junction. A similar dynamics can be observed for the tank with capacitance $C_2$ that results in the bond graph of Figure 5.11.

The procedure of assigning the causality to the bond graph starts by applying the required causality to the sources ($p_1$ and $q_1$), and the preferred causality to the storage devices (all $C$ and $L$ elements).

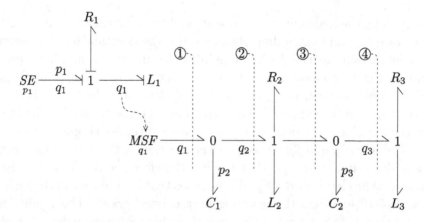

Figure 5.11  The bond graph of Problem 5.2.

The causality not yet assigned to the remaining components can be found via the required causality of the system multi-ports. In this case, the causal strokes of the capacitances $C_1$ and $C_2$ that are connected to the 0 junctions, and the inertances $L_1$, $L_2$ and $L_3$ connected to the 1 junction give us the base to assign the remaining causal relations. After completing this procedure we obtain the bond graph of Figure 5.12.

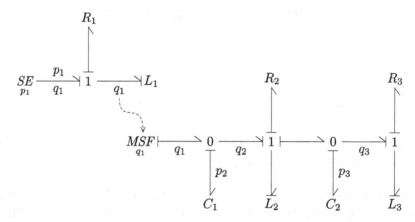

Figure 5.12  The bond graph of Problem 5.2 with all causality assigned.

The classical alternative dynamic equations of Problem 5.2 (considering zero initial conditions):

$$\begin{cases} q_1 = q_2 + q_{C_1}, \ q_2 = q_3 + q_{C_2} \\ p_1 = R_1 q_1 + p_{L_1}, \ p_2 = p_3 + R_2 q_2 + p_{L_2}, \ p_3 = R_3 q_3 + p_{L_3} \\ q_1 = \frac{1}{L_1} \int_0^t p_{L_1}(\tau) \, d\tau \\ q_2 = \frac{1}{L_2} \int_0^t p_{L_2}(\tau) \, d\tau \\ q_3 = \frac{1}{L_3} \int_0^t p_{L_3}(\tau) \, d\tau \\ p_2 = \frac{1}{C_1} \int_0^t q_{C_1}(\tau) \, d\tau \\ p_3 = \frac{1}{C_2} \int_0^t q_{C_2}(\tau) \, d\tau \end{cases}$$

The standard block diagram of the hydraulic system is shown in Figure 5.13.

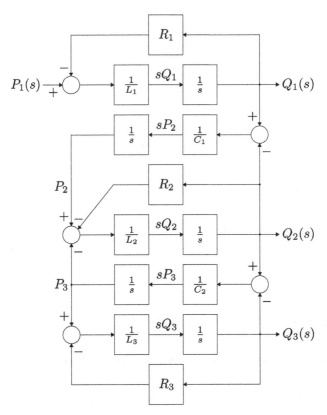

Figure 5.13 The block diagram of Problem 5.2 in the Laplace domain.

## 5.3 PROPOSED EXERCISES

**Exercise 5.1.** Consider the hydraulic system depicted in Figure 5.14 where the symbols $C_i$ ($i = 1, 2$), $R$, $p_i$ and $q_j$ ($j = 0, 1, 2$) represent the capacitances of the tanks, the hydraulic resistance, the pressures and the volume flow rates, respectively. The symbols $\alpha$, $\beta$, $\gamma$, $\delta$ and $\varepsilon$ included in the bond graph of Figure 5.15 correspond to:

A) $\alpha = SF$, $\beta = C_1$, $\gamma = C_2$, $\delta = SF$, $\varepsilon = R$

B) $\alpha = C_1$, $\beta = SF$, $\gamma = C_2$, $\delta = SF$, $\varepsilon = R$

C) $\alpha = C_1$, $\beta = SE$, $\gamma = C_2$, $\delta = SE$, $\varepsilon = R$

D) $\alpha = C_2$, $\beta = SF$, $\gamma = C_1$, $\delta = SF$, $\varepsilon = R$

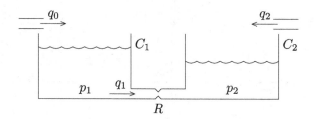

Figure 5.14  The hydraulic system of Exercise 5.1.

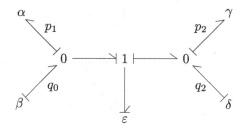

Figure 5.15  The bond graph of Exercise 5.1.

**Exercise 5.2.** Consider the hydraulic system shown in Figure 5.16(a) where $C_1$ and $C_2$ denote the capacitances of the tanks, $R_0$, $R_1$ and $R_2$ are hydraulic resistances, $q_i$, $q_0$, $q_1$ and $q_2$ represent the volume flow rates and, $p_1$ and $p_2$ stand for the pressures. Figure 5.16(b) shows the corresponding bond graph, where the symbols $\alpha$, $\beta$, $\gamma$ and $\varepsilon$ represent elements of the system. It can be concluded that the symbol $\alpha$ in the bond graph corresponds to:

**A)** $\alpha = C_1$

**B)** $\alpha = R_0$

**C)** $\alpha = R_1$

**D)** $\alpha = C_2$

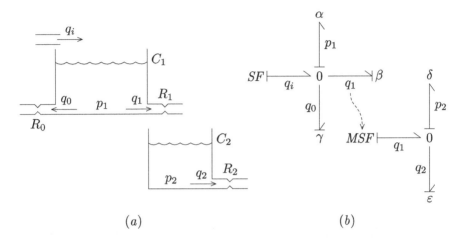

Figure 5.16  The (a) hydraulic system and (b) bond graph of Exercise 5.2.

**Exercise 5.3.** Consider the hydraulic system shown in Figure 5.17 where the symbols $C_i$ ($i = 1, 2, 3$) and $R_j$ ($j = 1, 2, 3, 4$) represent the capacitances of the tanks and the hydraulic resistances, respectively. Moreover, the symbols $p_i$ and $q_k$ ($k = 0, 1, 2, 3, 4$), stand for the pressures and the volume flow rates, respectively. The corresponding bond graph is given in Figure 5.18, where the symbols $\alpha$, $\beta$, $\gamma$, $\delta$ and $\varepsilon$ denote elements of the system. Then we can write:

**A)** $\alpha = R_1$, $\beta = C_3$, $\gamma = R_4$

**B)** $\alpha = R_1$, $\beta = C_2$, $\gamma = R_4$

**C)** $\alpha = C_1$, $\beta = C_2$, $\varepsilon = C_3$

**D)** $\alpha = C_1$, $\beta = C_2$, $\gamma = C_3$

**116** ■ An Introduction to Bond Graph Modeling with Applications

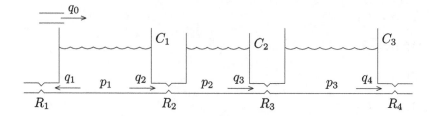

Figure 5.17 The hydraulic system of Exercise 5.3.

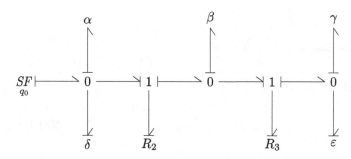

Figure 5.18 The bond graph of Exercise 5.3.

**Exercise 5.4.** Consider the hydraulic system shown in Figure 5.19 where the symbols $C_i$ ($i = 1, 2$), $R_i$ and $L$ stand for the capacitances of the tanks, the hydraulic resistances and the inertance, respectively. Moreover, the symbols $p_i$ and $q_j$ ($j = 0, 1, 2$), denote the pressures and the volume flow rates, respectively. Figure 5.20 shows the corresponding bond graph where the symbols $\alpha$, $\beta$ and $\gamma$ represent junctions of the bond graph. In this case:

**A)** $\alpha = 0$ and $\beta = 0$

**B)** $\beta = 0$ and $\gamma = 0$

**C)** $\alpha = 0$ and $\gamma = 0$

**D)** $\alpha = 1$ and $\beta = 0$

**Exercise 5.5.** Consider the hydraulic system shown in Figure 5.21. The symbols $C_i$ ($i = 1, 2$), $R_i$ and $L_i$ represent the capacitances of the tanks, hydraulic resistances and inertances, respectively. Additionally, the symbols $q_j$ ($i = 1, 2, 3, 4$) and $p_j$, stand for the volume flow rates and pressures, respectively. Draw the corresponding bond graph and discuss the causality.

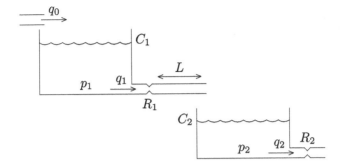

**Figure 5.19** The hydraulic system of Exercise 5.4.

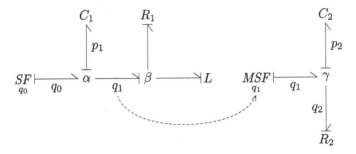

**Figure 5.20** The bond graph of Exercise 5.4.

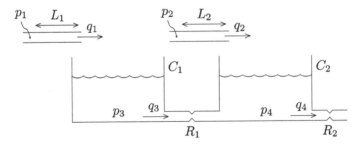

**Figure 5.21** The hydraulic system of Exercise 5.5.

**Exercise 5.6.** Consider the hydraulic system shown in Figure 5.22 and the corresponding bond graph in Figure 5.23. Let the symbols $C_i$ ($i = 1, 2$) and $R_i$ represent the capacitances of the tanks and the hydraulic resistances, respectively. Furthermore, the symbols $p_i$ and $q_j$ ($k = 0, 1, 2, 3$), stand for the pressures and the volume flow rates, respectively. The output flow of the pump is given by $q_3 = Kp_2$, where the symbol $K$ represents the gain of the device. The symbols $\alpha$, $\beta$, $\gamma$ and $\delta$ denote elements of the system. Then we can say:

**A)** $\alpha = R_1,\ \beta = C_1,\ \gamma = C_2,\ \delta = R_2$

**B)** $\alpha = C_1,\ \beta = R_1,\ \gamma = C_2,\ \delta = R_2$

**C)** $\alpha = R_1,\ \beta = C_1,\ \gamma = R_2,\ \delta = C_2$

**D)** $\alpha = C_1,\ \beta = R_1,\ \gamma = R_2,\ \delta = C_2$

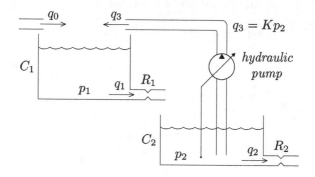

Figure 5.22  The hydraulic system of Exercise 5.6.

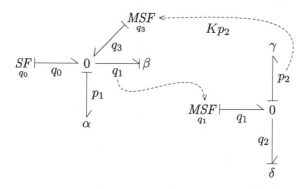

Figure 5.23  The bond graph of Exercise 5.6.

**Exercise 5.7.** Consider the hydraulic system presented in Figure 5.24, that comprises three tanks and the input flows $q_{i1}$, $q_{i2}$ and $q_{i3}$. The symbols $C_i$ ($i = 1, 2, 3$), $R_i$ and $L_i$ represent the capacitances of the tanks, the hydraulic resistances and the inertances, respectively. Additionally, the symbols $p_i$ and $q_i$, stand for the pressures and the volume flow rates, respectively. Figure 5.25 shows the corresponding bond graph with the causality assigned, where the symbols $\alpha$, $\beta$, $\gamma$ and $\delta$ denote elements of the system.

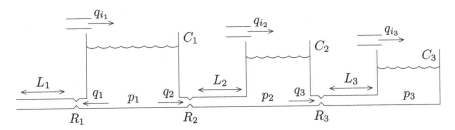

Figure 5.24  The hydraulic system of Exercise 5.7.

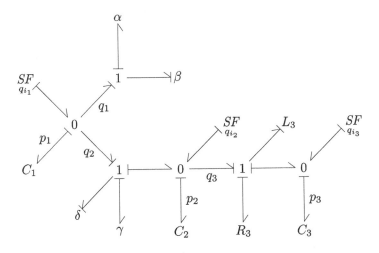

Figure 5.25  The bond graph of Exercise 5.7.

1. In the bond graph it can be stated that:

    **A)** $\alpha = R_2$
    **B)** $\beta = R_2$
    **C)** $\gamma = R_2$
    **D)** $\delta = R_2$

2. Sketch an analogous electrical circuit of the hydraulic system using the correspondence {pressure, flow rate}→{voltage, current}.

**Exercise 5.8.** Consider the hydraulic system depicted in Figure 5.26 that includes a variable displacement hydraulic pump, where the output flow is controlled by the proportional algorithm $q_0 = K_p(q_{ref} - q_3)$, where $K_p$ represents the controller gain. The flow $q_3$ is measured using a flow sensor connected to the output pipe. Moreover, are also represented the capacitances of the tanks $C_1$ and $C_2$, the hydraulic resistances $R_1$, $R_2$ and $R_3$ and the hydraulic inertances $L_1$, $L_2$ and $L_3$. The symbols $q_1$, $q_2$ and $q_3$ denote the volume flow rates and, $p_1$ and $p_2$ represent the pressures. Figure 5.27 shows the corresponding bond graph where the symbols $\alpha$, $\beta$, $\gamma$ and $\delta$ denote components of the system. In this case we have:

**A)** $\alpha = R_1$

**B)** $\beta = L_2$

**C)** $\gamma = L_1$

**D)** $\delta = R_2$

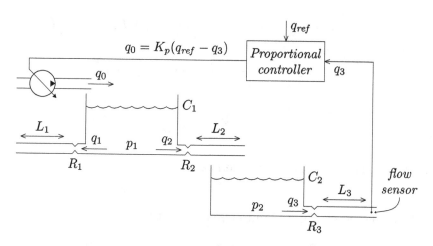

Figure 5.26 The hydraulic system of Exercise 5.8.

**Exercise 5.9.** Consider the hydraulic system shown in Figure 5.28, where $C_1$ and $C_2$ denote the capacitances of the tanks, $R_1$, $R_2$ and $R_3$ are hydraulic resistances and $L$ stands for the hydraulic inertance. The symbols $q_0$, $q_1$, $q_2$ and $q_3$ represent the volume flow rates and, $p_1$, $p_2$ and $p_3$ the pressures. Consider that the output flow of the pump is given by

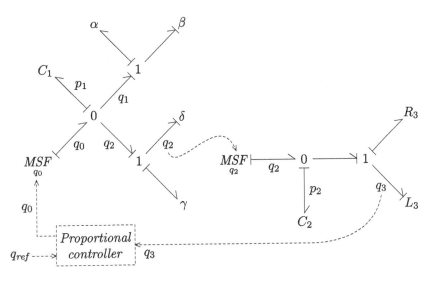

Figure 5.27 The bond graph of Exercise 5.8.

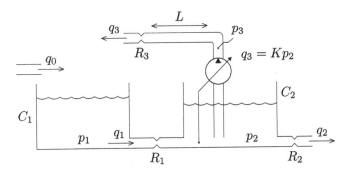

Figure 5.28 The hydraulic system of Exercise 5.9.

$q_3 = Kp_2$, where $K$ is the gain of the device. Find the bond graph and assign the causality.

**Exercise 5.10.** Consider the hydraulic system shown in Figure 5.29. The output flow of the pump is given by a proportional controller $q_0 = K(q_{ref} - q_2)$, where $K$ represents the controller gain. Let the symbols $C_i$, $R_i$ and $p_i$ ($i = 1, 2$) represent the capacitances of the tanks, hydraulic resistances and pressures, respectively. Additionally, the symbol $q_j$ ($j = 0, 1, 2$) denotes the fluid flows. Draw the corresponding bond graph and discuss the assigned causality.

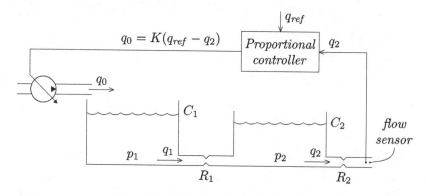

Figure 5.29  The hydraulic system of Exercise 5.10.

**Exercise 5.11.** Find a possible hydraulic system that can be modelled by the bond graph shown in Figure 5.30. Consider that the symbols $SE$, $SF$ and $MSF$ stand for effort source, flow source and modulated flow source, respectively. Moreover, the symbols $R_i$ ($i = 0, 1, 2, 3$), $C_1$ and $C_2$, and $L_0$ and $L_1$ represent the hydraulic resistances, capacitances and fluid inertances, respectively. Include all the effort and flow variables stated in the bond graph.

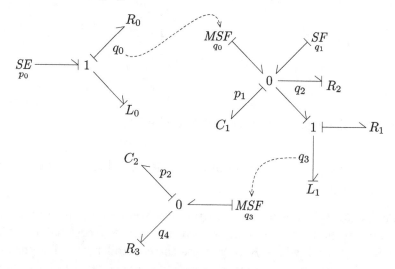

Figure 5.30  The bond graph of Exercise 5.11.

**Exercise 5.12.** Consider the hydraulic system shown in Figure 5.31 where the symbol $q_0$ denotes the input volume flow rate. The symbols $C_i$ ($i = 1, 2, 3$) and $R_j$ ($j = 1, 2$) represent the capacitances of the tanks and the hydraulic resistances, respectively. Moreover, the symbols $p_k$ ($k = 1, 2, 3, 4$) and $q_i$, stand for the pressures and the volume flow rates, respectively. The output flow of the pump is given by $q_2 = Kp_2$, $K > 0$, where the symbol $K$ represents the gain of the device. Find the bond graph and study the causality.

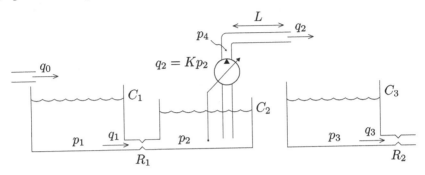

Figure 5.31 The hydraulic system of Exercise 5.12.

**Exercise 5.13.** Consider the hydraulic system shown in Figure 5.32, where the symbols $C_i$ ($i = 1, 2, 3$), $R_j$ ($i = 1, 2, 3, 4, 5$) and $L$ represent the capacitances of the tanks, hydraulic resistances and inertance, respectively. Additionally, the symbols $p_i$ and $q_k$ ($k = 0, 1, 2, 3, 4, 5$) denote the pressures and the volume flow rates, respectively. Draw the corresponding bond graph and assign the causality. If you find any causal conflict discuss possible ways for its solution.

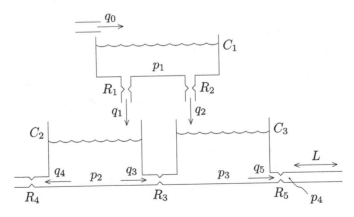

Figure 5.32 The hydraulic system of Exercise 5.13.

**Exercise 5.14.** Consider the hydraulic system shown in Figure 5.33, that comprises two input flows $q_{i1}$ and $q_{i2}$. Let the symbols $C_i$, $R_i$ and $L_i (i = 1, 2, 3)$ represent the capacitances of the tanks, the hydraulic resistances and inertances, respectively. Furthermore, the symbols $p_i$ and $q_i$, stand for the pressures and the volume flow rates, respectively.

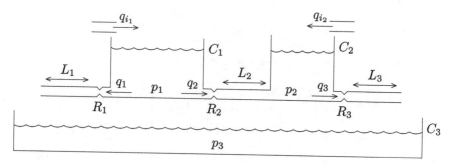

Figure 5.33 The hydraulic system of Exercise 5.14.

1. Write the equations that model the hydraulic system.
2. Find the bond graph of the system considering the causality.

**Exercise 5.15.** Consider the hydraulic system shown in Figure 5.34, where the output flow of the pump is controlled as follows $q_1 = Kp_1$. The symbol $K$ represents the gain of the device. Moreover, $C_i$ ($i = 1, 2, 3$), $R_i$ and $L$ are the capacitances of the tanks, the hydraulic resistances and the inertance, respectively. The symbols $q_j$ ($j = 0, 1, 2, 3, 4$) and $p_i$, stand for the volume flow rates and pressures, respectively. Draw the bond graph and discuss its causality.

**Exercise 5.16.** Consider the hydraulic system depicted in Figure 5.35. The input flows $q_1$ and $q_2$ are controlled by gate valves following the relations $q_1 = k_1 p_3$ and $q_2 = k_2 p_3$, where $k_1$ and $k_2$ are proportionally constants. Let $C_1$, $C_2$ and $C_3$ denote the capacitances of the tanks, $R_1$, $R_2$ and $R_3$ the hydraulic resistances, $L$ the hydraulic inertance, $q_i$ ($i = 3, 4, 5, 6$) the volume flow rates and, $p_1$, $p_2$ and $p_3$ the pressures. Draw the corresponding bond graph and discuss the causality.

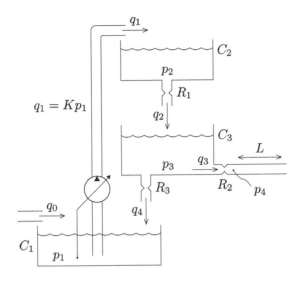

Figure 5.34 The hydraulic system of Exercise 5.15.

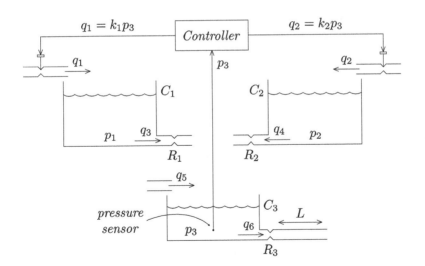

Figure 5.35 The hydraulic system of Exercise 5.16.

CHAPTER 6

# Thermal Systems

## 6.1 INTRODUCTION

Chapter 2 discussed the bond graph modeling. Several examples of thermal systems are analyzed in subsections 2.4.4 and 2.6.4. Often the temperature and the heat flow are the variables assigned as effort or flow. For constructing the bond graph, we can consider the following steps.

1. Find the system elements that have the same temperature and connect them by means of a 0 junction,

2. Find the system elements that share the same heat flow and connect them by means of a 1 junction,

3. Connect the 0 and 1 junctions to form a single bond graph,

4. Simplify the bond graph by eliminating any unnecessary element or junction,

5. Assign causality to the bond graph.

The sources of effort and flow are the system imposed temperature and heat flow, respectively. A 1 junction (or a 0 junction) is associated with a given heat flow (or a temperature). The dissipator, and flow store elements consist of the thermal resistance and capacitance, respectively. In thermal systems there is no effort store. Table 6.1 summarizes the bond graph elements for thermal systems.

## 6.2 SOLVED PROBLEMS

**Problem 6.1.** Consider the thermal system shown in Figure 6.1 that represents a wall with three layers. The symbols $C$, $R$, $T_i$ ($i = 0, 1, 2$) and

## 128 ■ An Introduction to Bond Graph Modeling with Applications

TABLE 6.1  Bond graph elements for thermal systems.

| Element | | Bond Graph |
|---|---|---|
| Effort | temperature, $T$ | |
| Flow | heat flow rate, $q$ | |
| Effort source | temperature source, $T$ | $SE \xrightarrow{T}$ |
| Flow source | heat flow source, $q$ | $SF \vdash\xrightarrow{q}$ |
| Effort store (inductor) | no equivalent | |
| Fow store (thermal capacity) | $T = \frac{1}{C} \int_0^t q(\tau) \mathrm{d}\tau$ or $T = \frac{1}{C} H$ | $\vdash\!\!\longrightarrow C$ |
| Dissipator (Fourier law) | $T = Rq$ | $\longrightarrow R$ <br> $\vdash\!\!\longrightarrow R$ |

$q_i$ denote the thermal capacitance, the thermal resistance, the temperatures and the heat flows, respectively. Consider also that the material of the left and right layers is the same and that it has a thermal resistance $R$ and neglectable thermal capacity. The material of the middle layer has a neglectable thermal resistance and the thermal capacitance $C$. The temperatures $T_0$ and $T_2$ stand for the interior and exterior temperature, respectively. Find the corresponding bond graph and apply the causal rules in order to assign the causality.

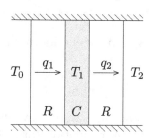

Figure 6.1  The thermal system of Problem 6.1.

**Resolution:** The bond graph for the conduction thermal system can be obtained as follows. The heat from the temperature source $T_0$ is passed to the wall material that has the thermal capacitance $C$, from which results the heat flow rate $q_1$. On the other hand, a temperature loss occurs due to conduction (Fourier's Law) when the flow of heat goes through the material with thermal resistance $R$. This dynamic relation introduces an effort (temperature) constraint between the temperature source $T_0$ and the (heat) flow store $C$, being therefore represented by a 1 junction. Figure 6.2 shows the bond graph with the 1 junction that connects the elements $SE$ and $R$ to the rest of the system.

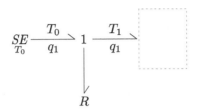

Figure 6.2 The partial bond graph of Problem 6.1.

The energy stored in the material with thermal capacitance $C$ is related to the difference between the heat inflow and outflow $(q_1 - q_2)$. This principle of thermal energy conservation is represented by a 0 junction where the sum of heat flow rates into a thermal junction must be zero. To connect the remaining elements we use the previous rationale given the symmetrical nature of the thermal system. Figure 6.3 shows the derived bond graph.

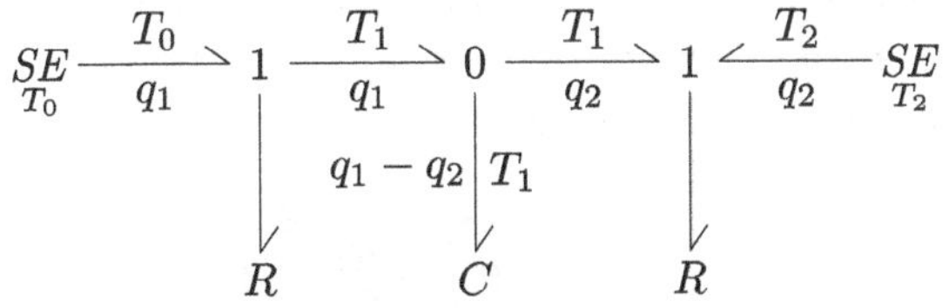

$$SE \xrightarrow[q_1]{T_0} 1 \xrightarrow[q_1]{T_1} 0 \xrightarrow[q_2]{T_1} 1 \xleftarrow[q_2]{T_2} SE$$
$$\underset{T_0}{\phantom{SE}} \quad \underset{R}{\downarrow} \quad \underset{C}{\overset{q_1-q_2 | T_1}{\downarrow}} \quad \underset{R}{\downarrow} \quad \underset{T_2}{\phantom{SE}}$$

Figure 6.3 The bond graph of Problem 6.1.

The causality assignment starts by establishing the required causality to the effort sources $T_0$ and $T_2$. Notice that following this step no further causal stokes can be assigned since both sources output effort and are connected to 1 junctions. In the second step the preferred causality is assigned to the storage element $C$ which results in the bond graph depicted in Figure 6.4.

The causality of the flow store element fulfills the requirements to assign the causality to the remaining bonds since it has effort as the output variable and is connected to a 0 junction. Using the causality rules of multi-port elements, the causal stokes of the thermal dissipators $R$ can be assigned. The result is the bond graph shown in Figure 6.5.

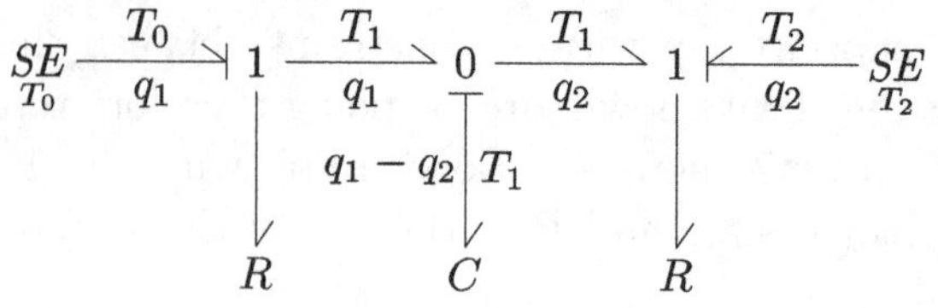

**130** ■ An Introduction to Bond Graph Modeling with Applications

$$SE_{T_0} \xrightarrow{T_0}_{q_1} 1 \xrightarrow{T_1}_{q_1} 0 \xrightarrow{T_1}_{q_2} 1 \xleftarrow{T_2}_{q_2} SE_{T_2}$$

$$\downarrow q_1 - q_2 \Big| T_1 \qquad \downarrow$$

$$R \qquad C \qquad R$$

Figure 6.4 The bond graph of Problem 6.1 after assigning causality to the sources and store.

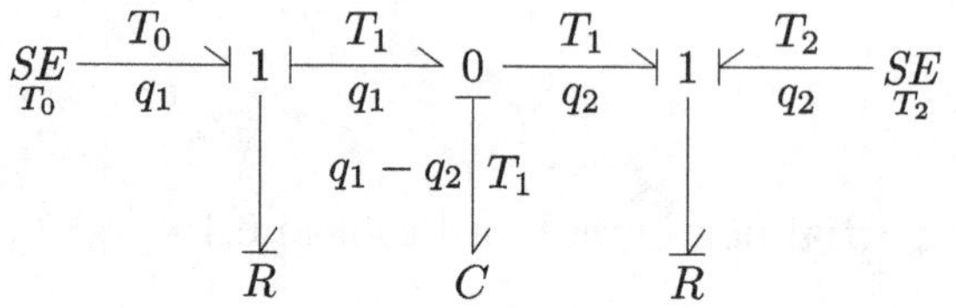

$$SE_{T_0} \xrightarrow{T_0}_{q_1} 1 \vdash \xrightarrow{T_1}_{q_1} 0 \xrightarrow{T_1}_{q_2} 1 \vdash \xleftarrow{T_2}_{q_2} SE_{T_2}$$

$$\downarrow q_1 - q_2 \Big| T_1 \qquad \downarrow$$

$$R \qquad C \qquad R$$

Figure 6.5 The bond graph of Problem 6.1 with the causality assigned.

**Problem 6.2.** Consider the thermal system shown in Figure 6.6 that includes three chambers and a inflow duct. Let the symbols $C_i$ ($i = 1, 2, 3$), $R_i$, $q_j$ ($j = 0, 1, 2, 3$) and $T_j$ stand for the thermal capacitances, the thermal resistances, the heat flows and the temperatures, respectively. Consider that there are no losses through the walls and that the temperature $T_0$ is constant. Draw the corresponding bond graph taking into account the causality of the elements.

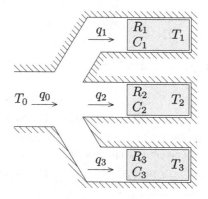

Figure 6.6 The thermal system of Problem 6.2.

**Resolution:** The procedure to find the bond graph model of Problem 6.2 can start by noticing that the temperature $T_0$ is the same across the inflow ducts. Given the different thermal materials located in the chambers and the geometry of the ducts, the main heat flow $q_0$ gets divided into three different heat flows ($q_1$, $q_2$ and $q_3$). This type of thermal interconnection constraint (first law of thermodynamics) is represented by means of a 0 junction, as shown in the resulting bond graph of Figure 6.7.

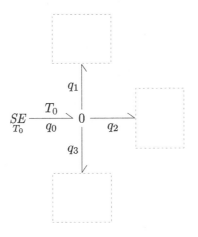

Figure 6.7  The partial bond graph of Problem 6.2.

The temperature source $T_0$ is in contact with the thermal flow stores $C_1$, $C_2$ and $C_3$ rising their temperatures ($T_1$, $T_2$ and $T_3$). In addition to the thermal capacitance, the chambers materials also have thermal resistance properties that cause a temperature drop due to conduction (Fourier's Law). As a result, we can model each chamber material using a 1 junction since the elements $C$, $R$, and the rest of the system to which they connect, all have different temperatures. This leads to the final bond graph in Figure 6.8.

The procedure for assigning the causality to the bond graph of Figure 6.8 is as follows:

1. The first step is to assign the required causality to the effort source $T_0$.

2. Follow with the assignment of the preferred causality to the storage elements: $C_1$, $C_2$ and $C_3$. These elements impose effort (use Table 6.1 for reference).

3. For each junction with an element with required or preferred causality, that meets the multi-port causal constraints, propagate the causality assignment through the graph. In this case only the

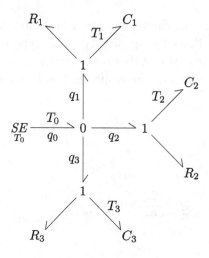

Figure 6.8 The bond graph of Problem 6.2.

effort source $T_0$ fulfils this criteria, which consequently enables the assignment of the causal strokes to the remaining bonds.

These steps result in the bond graph shown in Figure 6.9 that has no causality conflicts.

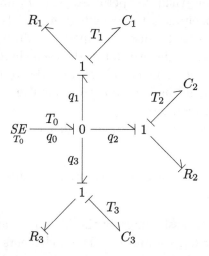

Figure 6.9 The bond graph of Problem 6.2 with the causality assigned.

If we adopt the standard model we obtain the set of equations (with zero initial conditions):

$$\begin{cases} q_0 = q_1 + q_2 + q_3 \\ T_0 - T_1 = R_1 q_1 \\ T_0 - T_2 = R_2 q_2 \\ T_0 - T_3 = R_3 q_3 \\ T_1 = \frac{1}{C_1} \int_0^t q_1(\tau) \, d\tau \\ T_2 = \frac{1}{C_2} \int_0^t q_2(\tau) \, d\tau \\ T_3 = \frac{1}{C_3} \int_0^t q_3(\tau) \, d\tau \end{cases}$$

The standard block diagram is represented in Figure 6.10. We verify that the bond graph modeling leads to a much more intuitive and clear representation.

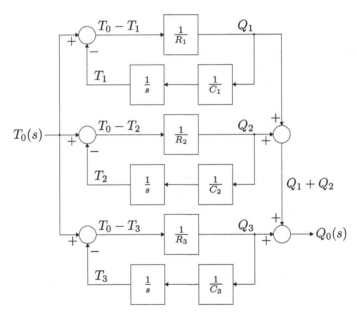

Figure 6.10 The block diagram of Problem 6.2 in the Laplace domain.

Alternatively, we can consider the heat flow $q_0$ as the energy source instead of the temperature $T_0$. This approach leads to the 'indeterminate' bond graph shown in Figure 6.11 where, in order to complete the causal assignment, either of the three thermal resistances must be granted with an arbitrary causality.

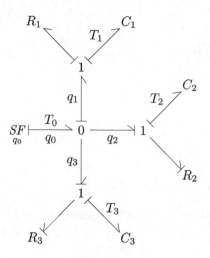

Figure 6.11  The alternative bond graph of Problem 6.2.

## 6.3 PROPOSED EXERCISES

**Exercise 6.1.** Consider the thermal system shown in Figure 6.12(a) that represents a wall with three distinct materials and a source of heat flow (produced by means of an electrical resistance). The symbols $C_i$ ($i = 1, 2$), $R$ and $q_i$ stand for the thermal capacitances, the thermal resistance and the heat flows, respectively. The material of the middle layer has a neglectable thermal capacity and the thermal resistance $R$. Figure 6.12(b) shows the corresponding bond graph where the symbols $\alpha$, $\beta$, $\gamma$ and $\delta$ denote elements of the bond graph.

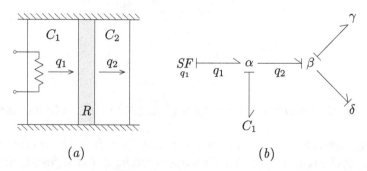

Figure 6.12  The (a) thermal system and (b) bond graph of Exercise 6.1.

In the bond graph we can say:

**A)** $\alpha = 0$, $\beta = 1$, $\gamma = C_2$, $\delta = R$

**B)** $\alpha = 0$, $\beta = 1$, $\gamma = R$, $\delta = C_2$

**C)** $\alpha = 1$, $\beta = 0$, $\gamma = R$, $\delta = C_2$

**D)** $\alpha = 1$, $\beta = 0$, $\gamma = C_2$, $\delta = R$

**Exercise 6.2.** Consider the thermal system shown in Figure 6.13(a), where the symbols $C_i$ ($i = 1, 2$), $R_i$ and $T_j$ ($j = 0, 1, 2$) represent thermal capacitances, thermal resistances and temperatures, respectively. The bond graph of the system is also sketched in Figure 6.13(b), where the symbols $\alpha$, $\beta$, $\gamma$, $\delta$ and $\varepsilon$ denote elements of the system. Then we can say that the symbols:

**A)** $\alpha = 0$, $\beta = 1$, $\gamma = 0$

**B)** $\alpha = 1$, $\beta = 0$, $\delta = R_2$

**C)** $\alpha = 1$, $\beta = 0$, $\varepsilon = R_2$

**D)** $\gamma = 1$, $\delta = R_2$, $\varepsilon = C_2$

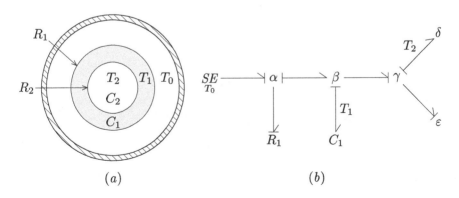

(a)          (b)

**Figure 6.13** The (a) thermal system and (b) bond graph of Exercise 6.2.

**Exercise 6.3.** Consider the thermal system shown in Figure 6.14 having two chambers and a heater, where the symbols $C_i$ ($i = 1, 2, 3$) and $R_i$ represent the thermal capacitances and the thermal resistances, respectively. Let the symbol $T_j$ ($j = 0, 1, 2, 3$) stand for the temperatures and the symbol $q_i$ denote the heat flows. Figure 6.15 shows the corresponding bond graph, where the symbols $\alpha$, $\beta$, $\gamma$, $\delta$ and $\varepsilon$ are elements of the system. Then we can say:

**A)** $\alpha = C_1$, $\beta = C_2$

**B)** $\gamma = C_2$, $\delta = C_3$

**C)** $\alpha = C_2$, $\varepsilon = C_3$

**D)** $\beta = C_2$, $\gamma = C_1$

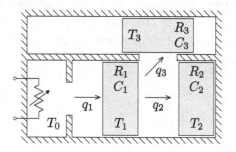

Figure 6.14  The thermal system of Exercise 6.3.

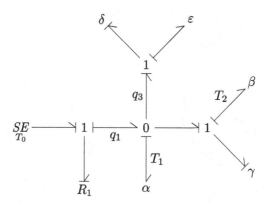

Figure 6.15  The bond graph of Exercise 6.3.

**Exercise 6.4.** Consider the thermal system shown in Figure 6.16, where the symbols $T$, $R$ and $C$ denote the temperatures, the thermal resistances and the thermal capacitances, respectively. There are five materials with different thermal properties called air, powder, fluid, glass and mercury, represented by the symbols 'a', 'p', 'f', 'g' and 'm', respectively. Consider that the air and powder temperatures, $T_a$ and $T_p$, are constant and that the powder container is perfectly insulated to the outside. The system has five temperatures $\{T_p, T_f, T_g, T_m, T_a\}$, four thermal resistances $\{R_{pf}, R_{fg}, R_{gm}, R_{ma}\}$ and three thermal capacitances $\{C_f, C_g, C_m\}$. The symbols $\alpha$, $\beta$, $\gamma$, $\delta$ and $\varepsilon$ included in the bond graph of Figure 6.17 correspond to:

**A)** $\alpha = R_{fg}$, $\gamma = C_g$, $\delta = R_{pf}$

**B)** $\beta = R_{pf}$, $\gamma = R_{fg}$, $\delta = B_3$

**C)** $\gamma = R_{fg}$, $\delta = R_{ma}$, $\varepsilon = R_{gm}$

**D)** $\alpha = R_{pf}$, $\gamma = C_g$, $\varepsilon = R_{ma}$

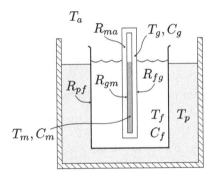

Figure 6.16 The thermal system of Exercise 6.4.

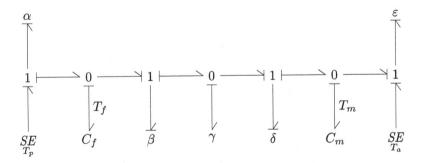

Figure 6.17 The bond graph of Exercise 6.4.

**Exercise 6.5.** Consider the bond graph in Figure 6.18 where the symbols $SE$, $R$ and $C$ denote the effort source, the dissipator elements and the flow stores, respectively.

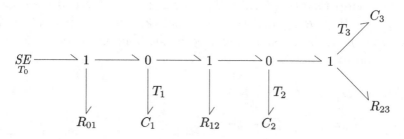

Figure 6.18 The bond graph of Exercise 6.5.

1. Assign causality to the bond graph elements.

2. Sketch a thermal system that can be modeled by the bond graph in Figure 6.18. Consider that the system includes four materials (0, 1, 2 and 3) with different thermal properties.

**Exercise 6.6.** Consider the thermal system shown in Figure 6.19, where the symbols $C_i$ ($i = 1, 2$), $R_i$, $q_i$ and $T_j$ ($j = 0, 1, 2$) denote the thermal capacitances, the thermal resistances, the heat flows and the temperatures, respectively. Find the bond graph considering the causality between the elements.

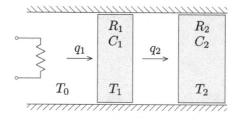

Figure 6.19 The thermal system of Exercise 6.6.

**Exercise 6.7.** Consider the thermal system shown in Figure 6.20 that comprises four materials with different thermal properties called air, fluid, glass and mercury, represented by the symbols '$a$', '$f$', '$g$' and '$m$', respectively. Let the symbols $T$, $R$, and $C$ stand for the temperatures, the thermal resistances, and the thermal capacitances, respectively. Consider that the air temperature $T_a$ is constant and that the system has four temperatures $\{T_f, T_g, T_m, T_a\}$, four thermal resistances $\{R_{fg}, R_{gm}, R_{fa}, R_{ga}\}$ and three thermal capacitances $\{C_f, C_g, C_m\}$. Draw the bond graph of the system considering the causality of the components.

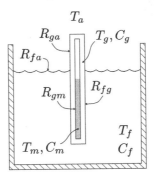

Figure 6.20  The thermal system of Exercise 6.7.

**Exercise 6.8.** Figure 6.21 shows a thermal system that has paths involving three distinct materials and a source of heat flow produced by means of an electrical resistance. Consider that the symbols $C_i$ ($i = 1, 2, 3$), $R_i$, $q_i$ and $T_j$ ($j = 0, 1, 2, 3$) represent the thermal capacitances, the thermal resistances, the heat flows and the temperatures, respectively. Find the bond graph of the system and assign the causality.

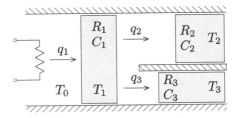

Figure 6.21  The thermal system of Exercise 6.8.

**Exercise 6.9.** Figure 6.22 shows a section cut of a thermal system that has a cylindrical shape and three different materials represented by the indexes 0, 1 and 2. Let the symbols $T$, $R$ and $C$ denote the temperature, the thermal resistance and the thermal capacitance, respectively. Consider that the system has three temperatures $\{T_0, T_1, T_2\}$, three thermal resistances $\{R_{01}, R_{02}, R_{12}\}$ and two thermal capacitances $\{C_1, C_2\}$. Moreover, consider that the temperature $T_0$ is constant. Find the bond graph of the system and assign the causality.

**140** ■ An Introduction to Bond Graph Modeling with Applications

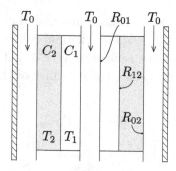

Figure 6.22 The thermal system of Exercise 6.11.

**Exercise 6.10.** Consider the thermal system shown in Figure 6.23, where the symbols $C_i$ ($i = 3, 4, 5, 6$), $R_j$ ($j = 1, \ldots, 6$), $T_k$ ($k = 0, \ldots, 6$) and $q_l$ ($l = 0, \ldots, 7$) are the thermal capacitances, the thermal resistances, the temperatures and heat flows, respectively. Take into consideration that there are no losses through the walls and that the temperature $T_0$ is constant.

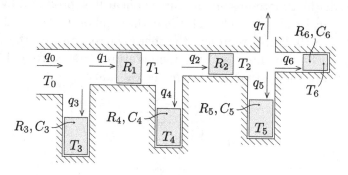

Figure 6.23 The thermal system of Exercise 6.10.

1. Draw the bond graph of the system and assign the causal strokes.
2. Find the mathematical model and the corresponding block diagram of the thermal system.

**Exercise 6.11.** Consider the thermal system shown in Figure 6.24 that includes several chambers, where two of them have sources of heat flow implemented by means of electrical resistances. Let the symbols $C_i$ ($i = 1, 2$), $R_j$ ($j = 1, 2, 3$), $q_j$ and $T_k$ ($k = 1, 2, a, b$) stand for the thermal capacitances, the thermal resistances, the heat flows and the temperatures, respectively. Consider that there are no losses through the walls and that the resistances are regulated so that the temperatures $T_a$ and $T_b$ are constant. Draw the corresponding bond graph taking into account the causality of the elements.

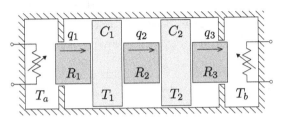

Figure 6.24  The thermal system of Exercise 6.11.

**Exercise 6.12.** Consider the thermal system of Figure 6.25 that has paths involving four distinct materials and two sources of heat flow implemented by means of electrical resistances. Let the symbols $C_i$ ($i = 1, 2, 4$), $R_j$ ($j = 1, 2, 3, 4$), $q_k$ ($j = 0, 1, 2, 3, 4$) and $T_l$ ($l = 1, 2, 3, 4, a, b$) denote the thermal capacitances, the thermal resistances, the heat flows and the temperatures, respectively. Consider that there are no losses through the walls and that the temperatures $T_a$ and $T_b$ are constant.

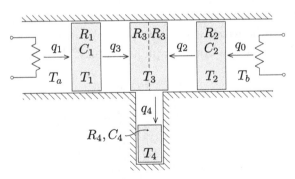

Figure 6.25  The thermal system of Exercise 6.12.

1. Obtain the bond graph of the system and assign the causality.

2. Sketch an analogous electrical circuit of the thermal system using the {temperature, heat flow rate}→{voltage, current} analogy.

**Exercise 6.13.** Consider the thermal system shown in Figure 6.26 that comprises two chambers, both having heat flow sources by means of electrical resistances. Let the symbols $C_i$ $(i = 1, 2, 3)$, $R_j$ $(j = 1, \ldots, 6)$, $q_j$ and $T_k$ $(i = 1, 2, 3, a, b)$ represent thermal capacitances, thermal resistances, heat flows and temperatures, respectively. We have no thermal losses through the walls and we consider that the resistances are controlled so that the temperatures $T_a$ and $T_b$ are constant. Draw the corresponding bond graph and discuss the causality.

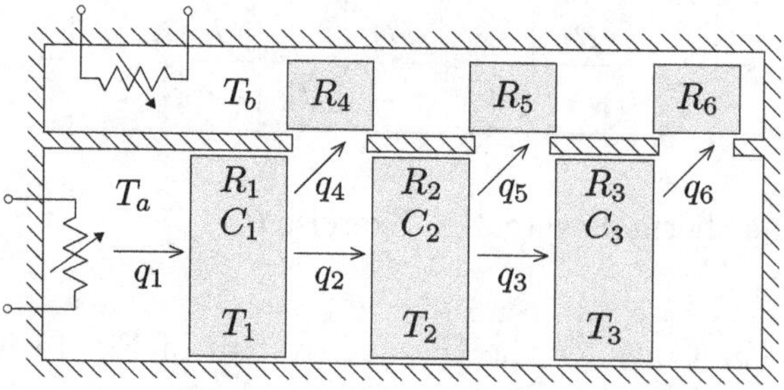

Figure 6.26 The thermal system of Exercise 6.13.

**Exercise 6.14.** Consider the thermal system shown in Figure 6.27, where the three electrical resistances are controlled so that the temperatures $T_1$, $T_2$ and $T_3$ are kept constant. The system includes materials with thermal resistances $R_i$ $(i = 1, 2, 3, 4)$ and, in the center, a material at temperature $T$ and with thermal capacitance $C$. In addition, there is a heat flow $q$ to the exterior (at zero temperature) through the thermal resistance $R_4$. Draw the bond graph and assign the causality.

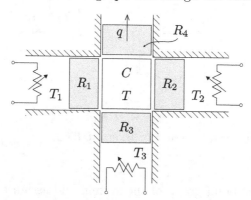

Figure 6.27 The thermal system of Exercise 6.14.

CHAPTER 7

# Multi-domain Systems

## 7.1 INTRODUCTION

In the previous chapters we discussed the representation of dynamical systems using bond graphs and illustrated the concepts in the scope of different types of systems, often called as 'domains'. Nonetheless, one of the key characteristics of this modeling technique is that it applies straightforwardly to systems involving different energy domains. Indeed, often bond graphs are said to be a multi-domain modeling approach.

## 7.2 SOLVED PROBLEMS

**Problem 7.1.** Consider the electromechanical system shown in Figure 7.1. To simplify the analysis consider that both the force $f$, produced by the solenoid, and the electromotive force $e_b$, induced by the ferromagnetic core, have approximately linear relations given by the expressions $f = K_i i_a$ and $e_b = K_i \dot{x}_1$, respectively. Let $K_i$ be a proportionality constant, $i_a$ the electric current in the solenoid and $\dot{x}_1$ the velocity of the core with mass $M_1$. Draw the bond graph and discuss the causality.

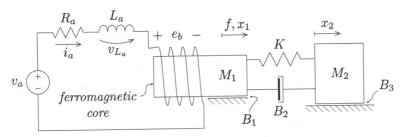

Figure 7.1 The electromechanical system of Exercise 7.1.

**Resolution:** To obtain the bond graph representation from the ideal model of Figure 7.1 we first identify that it covers two energy domains, the electrical and (linear) mechanical, coupled by means of a domain-transformation transducer. Based on this analysis we first derive the partial bond graph of Figure 7.2(a) that refers to the electrical part of the system. This bond graph models the fact that all electrical components share the same current $i_a$, being therefore, connected by means of a 1 junction. The bond with the effort variable $e_b$ models the presence of the electromotive force in the series loop, alongside the voltage drops $v_a$, $R_a i_a$ and $v_{L_a}$.

In mechanical systems modeling, elements that share the same velocity are connected using a 0 junction. In this case we have three pairs of elements: $M_1$ and $B_1$, $M_2$ and $B_3$, and $K$ and $B_2$, that share the linear velocities $\dot{x}_1$, $\dot{x}_2$ and $\dot{x}_1 - \dot{x}_2$, respectively. However, these three pairs of elements are physically linked and have different velocities between them, together with a commonly shared force. Hence, the three 0 junctions are interconnected by a 1 junction as shown in the partial bond graph of Figure 7.2(b). If we consider that the energy source of the mechanical subsystem is the force $f$ produced by the solenoid, it can be connected to the 0 junction on the left.

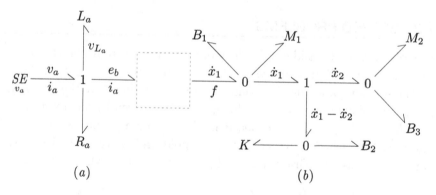

Figure 7.2 The (a) electrical and (b) linear mechanical subsystems bond graphs of Problem 7.1.

The coupling electromagnetic device is governed by the relations $f = K_i i_a$ and $e_b = K_i \dot{x}_1$, that comply with the bond graph transformer element equations (see subsection 2.3.4). Therefore, Figure 7.3 shows the complete bond graph attained from the previous two partial bond graphs interconnected using the $TF$ element with the transformation ratio $\frac{1}{K_i}$.

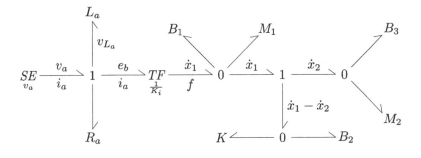

**Figure 7.3** The bond graph of Problem 7.1 without the causality assigned.

The procedure to assign the causal strokes to the bond graph of Figure 7.3 is as follows:

1. First we assign the required causality to the voltage source $v_a$.

2. The second step is to assign the integral causality to the storage devices, in this case the inductor $L_a$, the masses $M_1$ and $M_2$, and the spring $K$.

3. The next step propagates causality based on the elements that satisfy the junctions causal constraints: the stores $L_a$, $M_1$ and $M_2$. This step also accomplishes the causality assignment to the $TF$ component.

4. The causality not yet assigned is found via the causal rules of the mechanical subsystem 1 junction.

This sequence gives us the bond graph shown in Figure 7.4 that has no causal conflict.

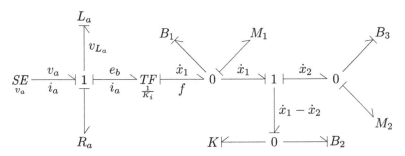

**Figure 7.4** The complete bond graph of Problem 7.1.

The conventional classical dynamic equations of Problem 7.1 (considering zero initial conditions):

$$\begin{cases} v_a = R_a i_a + v_{L_a} + e_b, \ i_a = \frac{1}{L_a}\int_0^t v_{L_a}(\tau)\,d\tau \\ e_b = K_i \dot{x}_1, \ f = K_i i_a \\ f = M_1 \ddot{x}_1 + B_1 \dot{x}_1 + K(x_1 - x_2) + B_2(\dot{x}_1 - \dot{x}_2) \\ M_2 \ddot{x}_2 + B_3 \dot{x}_2 = K(x_1 - x_2) + B_2(\dot{x}_1 - \dot{x}_2) \end{cases}$$

The standard block diagram of the system is depicted in Figure 7.5. We observe that the bond graph modeling tool results in a much more concise representation.

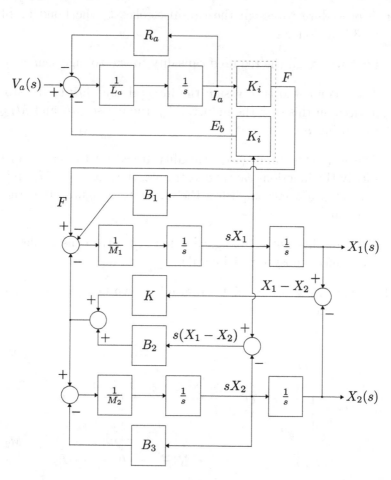

Figure 7.5 The block diagram of Problem 7.1 in the Laplace domain.

**Problem 7.2.** Consider the multi-domain system shown in Figure 7.6 where the DC motor drives the (fixed displacement) hydraulic pump through a rigid shaft with torque $T$ and angular velocity $\omega$. The hydraulic circuit feeds the cylinder whose rod acts with force $f$ on the mechanical load composed by the mass $M$, the spring with stiffness constant $K$ and the damper with viscous friction coefficient $B$. In the hydraulic subsystem $p$ and $p_c$ denote pressures, $q$ represents the volume flow rate and $R_h$ the hydraulic resistance. On the electrical part of the system, $v_a$ is the applied voltage to the DC motor armature, $i_a$ the current passing though it and $e_b$ the induced electromotive force. Moreover, consider the following relations:

- $e_b = K_b \omega$ and $T = K_b i_a$, where $K_b$ is the DC motor constant;

- $q = K_h \omega$ and $p = \frac{1}{K_h} T$, where $K_h$ is a constant related to the pump's capacity;

- $p_c = \frac{f}{A}$ and $q = A\dot{x}$, where $A$ is the hydraulic cylinder piston area and $\dot{x}$ the linear velocity of the mechanical load.

Find the bond graph representation of the system and discuss the causality.

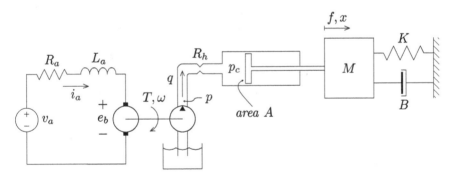

Figure 7.6 The multi-domain system of Problem 7.2.

**Resolution:** Figure 7.7 lays out the energy domains analysis taken into account for the bond graph construction. The system includes the electrical, mechanical (linear and rotational) and hydraulic domains which are coupled together using three power transducer in the form of a DC motor, hydraulic pump and hydraulic cylinder. Notice that the rotational mechanical subsystem has no sources, stores or dissipators.

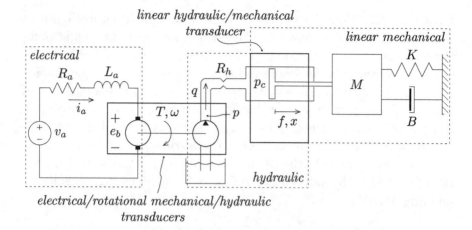

Figure 7.7 Analysis of the multi-domain system of Problem 7.2.

In the purely electrical subsystem all elements are in series, therefore, they are connected by means of a 1 junction, as seen in Figure 7.8($a$). The output of the hydraulic pump provides the flow rate $q$ that goes through $R_h$, and into the hydraulic cylinder chamber. The resistance $R_h$ is then connected using a 1 junction since it is subject to flow $q$ and the pressure drop $p - p_c$ (Figure 7.8($b$)). All linear mechanical elements share the same velocity $\dot{x}$, as a result, the corresponding partial bond graph of Figure 7.8($c$) shows the elements connected by a 0 junction.

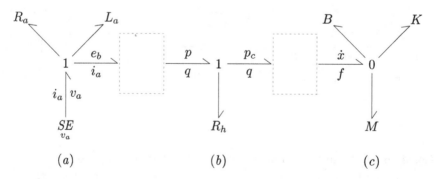

Figure 7.8 The ($a$) electrical, ($b$) hydraulic and ($c$) linear mechanical subsystems bond graphs of Problem 7.2.

We proceed to determine if the transducers are modeled by the $TF$ or $GY$ bond graph elements. Consider the relations that describe these power conservative two-ports elements (refer to subsection 2.3.4):

- TF: $e_2 = n \cdot e_1$, $f_1 = n \cdot f_2$, where $n$ is the transformer ratio,
- GY: $e_1 = r \cdot f_2$, $e_2 = r \cdot f_1$, where $r$ is the gyration resistance.

For the three system transducers we find that:

- DC motor: $\omega = \frac{1}{K_b} \cdot e_b$ and $i_a = \frac{1}{K_b} \cdot T$ comply with the $TF$ relations, where $\frac{1}{K_b} = n$ is the transformation ratio;

- Hydraulic pump: $p = \frac{1}{K_h} \cdot T$ and $\omega = \frac{1}{K_h} \cdot q$ match the $GY$ relations, where $\frac{1}{K_h} = r_1$ is the gyration resistance;

- Hydraulic cylinder: $p_c = \frac{1}{A} \cdot f$ and $\dot{x} = \frac{1}{A} \cdot q$ conform to the $GY$ relations. Consider $\frac{1}{A} = r_2$;

Based on this information the complete bond graph is shown in the Figure 7.9.

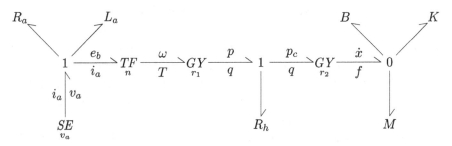

Figure 7.9 The bond graph of Problem 7.2 without the causality assigned, where $n = \frac{1}{K_b}$, $r_1 = \frac{1}{K_h}$ and $r_2 = \frac{1}{A}$.

Applying the rules described in section 2.6, the causality assignment to the bond graph of Figure 7.9 is as follows:

1. Assign the prescribed causality to the only system source, $v_a$.

2. The second rule is to assign the preferred causality to stores, in this case, the $L_a$, $M$ and $K$ elements.

3. Next we follow the consequences of the previous steps by assigning causality to the other system elements by means of the system two ports and junctions causal rules. Consider that with the $TF$ element the causality of the output bond is the same as of the input bond, and that for the $GY$ element the causality of the input and output bonds are the opposite.

This procedure leads to the bond graph shown in Figure 7.10 that has no causal conflict.

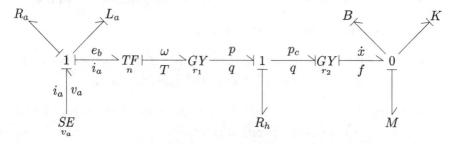

Figure 7.10  The complete bond graph of Problem 7.2.

## 7.3  PROPOSED EXERCISES

**Exercise 7.1.** A loudspeaker is a device that converts electrical signals into sound as the result of an electromechanical process, such device is illustrated in Figure 7.11. The voltage $v_i$ causes an electric current $i$ that, when passing through the coil, produces a vibration $x$. In the electrical part of the loudspeaker an induced electromotive force $e = K_e \dot{x}$ emerges, while on the mechanical part a force $f_m = K_e i$ is generated. The mechanical subsystem comprises the elements mass, spring and damper represented in the figure by the symbols $M$, $K$ and $B$, respectively. Figure 7.12 shows the corresponding bond graph where the symbols $\alpha$, $\beta$ and $\delta$ represent elements of the system.

1. In the bond graph we have:

    **A)** $\alpha = M$, $\beta = B$, $\delta = K$
    **B)** $\alpha = K$, $\beta = B$, $\delta = M$
    **C)** $\alpha = B$, $\beta = M$, $\delta = K$
    **D)** $\alpha = M$, $\beta = K$, $\delta = B$

2. The bond graph of Figure 7.12 has:

    **A)** Causal conflict in two connections
    **B)** No causal conflict
    **C)** Causal conflict in one connection
    **D)** Causal conflict in three connections

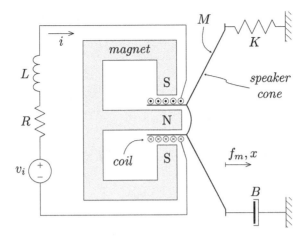

Figure 7.11 The electromechanical system of Exercise 7.1.

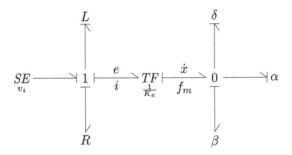

Figure 7.12 The bond graph of Exercise 7.1.

**Exercise 7.2.** Consider the mechanical transmission system illustrated in Figure 7.13, where the DC motor generated torque $T_m$ is applied to a load (with the elements $J_L$ and $B_L$) using a non rigid transmission shaft modeled by a friction $B$ and elasticity $K$. The armature controlled DC motor is characterized by the inductance $L_a$, resistance $R_a$, back-emf $e_b = K_b \dot{\theta}_m$, inertia $J_m$ and rotational friction $B_m$. Consider that the motor torque is given by $T_m = K_T i_a$, where $K_T = K_b$. The bond graph of the system is shown in Figure 7.14, where the symbols $\alpha$, $\beta$, $\gamma$, $\delta$ and $\varepsilon$ represent elements of the system. Then we can say:

**A)** $\alpha = 1$

**C)** $\alpha = TF$

**B)** $\alpha = 0$

**D)** $\alpha = GY$

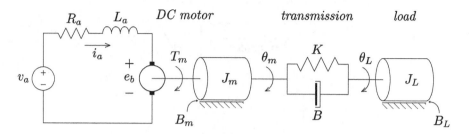

Figure 7.13 The electromechanical system of Exercise 7.2.

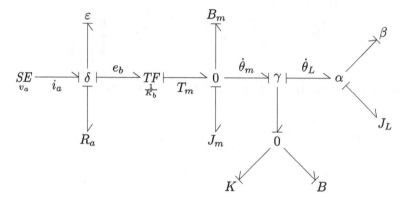

Figure 7.14 The bond graph of Exercise 7.2.

**Exercise 7.3.** Consider the electromechanical system shown in Figure 7.15 composed by a load (with inertia $J_L$ and linear friction $B_L$) driven by a armature controlled DC motor (with inductance $L_a$, resistance $R_a$, back-emf $e_b = K_b \dot{\theta}_m$, inertia $J_m$ and rotational friction $B_m$). The transmission is implemented by means of a lead screw with pitch $h$. Furthermore, the screw is not ideal and includes a friction coefficient $B$ and a stiffness constant $K$. The DC motor provides the torque $T_m = K_T i_a$ (consider $K_T = K_b$) for the angular displacement $\theta_m$ and linear displacement $x$ of mass $M_L$. Find the bond graph of the system and discuss the causality.

**Exercise 7.4.** Consider the electromechanical system shown in Figure 7.16 and the corresponding bond graph in Figure 7.17. The solenoid core is used to drive a rack and pinion system where the pinion rotates on a fixed axis and has radius $r$. The force $f_1$ produced by the solenoid and the electromotive force $e_b$, induced by the core, have approximately linear relations, given by the expressions $f_1 = K_i i$ and $e_b = K_i \dot{x}_1$, respectively. Let $K_i$ be a proportionality constant, $i$ the electric current in

Multi-domain Systems ■ 153

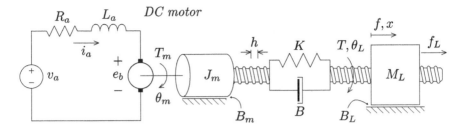

Figure 7.15 The electromechanical system of Exercise 7.3.

the solenoid and $\dot{x}$ the velocity of the core with mass $M_1$. The conversion from translational to rotational motion performed by the rack and pinion system is given by the relations $\dot{x} = r\omega_1$ and $rf_2 = T$, where the symbol $T$ denotes the torque and $\omega_1$ is the angular velocity. Additionally, in the figure is also represented the inertia $J_2$, the springs with stiffness constants $K_1$ and $K_2$, and the viscous friction coefficients $B_1$ and $B_2$.

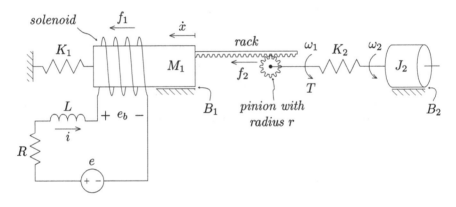

Figure 7.16 The electromechanical system of Exercise 7.4.

1. The symbols $\alpha$ and $\beta$ in the bond graph of Figure 7.17 represent:

    **A)** $\alpha = \omega_1,\ \beta = K_2$
    **B)** $\alpha = \omega_2,\ \beta = K_2$
    **C)** $\alpha = \omega_1,\ \beta = B_2$
    **D)** $\alpha = \omega_2,\ \beta = B_2$

2. Assign the causality to the bond graph.

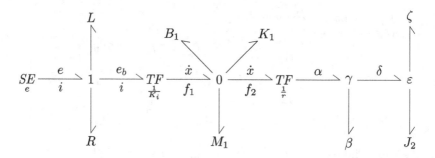

Figure 7.17 The bond graph of Exercise 7.4.

**Exercise 7.5.** The electromechanical system illustrated in Figure 7.18 represents a accelerometer sensor. The working principle is based on the inertial effect associated with the mass $M$ (Newton's second law) connected to the case through the spring $K$ (Hooke's law) and damper $B$. The displacement transducer is a potenciometer that provides the variable voltage $v_i$ to the amplification stage. This voltage is proportional to the mass $M$ relative displacement $y$ following the relation $v_i = A_1 y$, where $A_1$ is the transducer device gain. The active amplification stage uses a device with gain $A_2$ and very high input impedance $R_i$, that outputs voltage $v_2$, with $v_2 = A_2 v_i$.

Generally the potentiometric accelerometers have a low natural frequency, making them best suited for low frequency vibration measure-

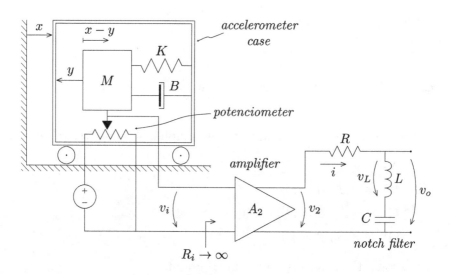

Figure 7.18 The accelerometer of Exercise 7.5.

ments or steady-state acceleration. As a result, to improve the quality of the output signal $v_o$ a notch filter is used to cutoff unwanted frequencies.

1. Draw the bond graph of the system considering the causality of the components.

2. Determine the mathematical model and the corresponding block diagram of the electromechanical system.

**Exercise 7.6.** The electromechanical system represented in the Figure 7.19 involves a capacitive microphone that converts mechanical to electrical energy. The plates $a$ and $b$ of the capacitor are fixed and movable respectively. Plate $b$ vibrates according to the sound waves and the voltage drop across the resistance $R$ reproduces approximately the sounds.

Consider that the plates have an electrical charge $q_0$ and that the attraction forces between the two plates are counterbalanced by the parallel association of spring and damper with coefficients $K$ and $B$, respectively. Each capacitor plate has area $A$ and the movable plate has mass $M$. Let $x_0$ and $x_1$ denote the distance between the capacitor plates and the length of the spring and dashpot, at steady state equilibrium, respectively. The symbols $E$ and $L$ represent the applied voltage and the inductance in the electrical circuit.

Consider that the sound waves produce a force $f(t)$ that leads to a small mechanical displacement $x(t)$ and electrical current and charge

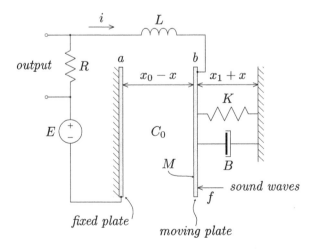

Figure 7.19 The electromechanical system of Exercise 7.6.

$i(t) = \frac{dq}{dt}$ and $q(t)$, respectively. The nonlinear model of this system is given by the two equations:
$$\begin{cases} M\ddot{x}(t) + B\dot{x}(t) - \frac{1}{2\epsilon A}(q_0 + q(t))^2 + K(x_1 + x(t)) = f(t) \\ L\ddot{q}(t) + R\dot{q}(t) + \frac{1}{\epsilon A}(q_0 + q(t))(x_0 - x(t)) = E \end{cases}$$
where $\epsilon$ stands for the electrical permittivity of the material between the two capacitor plates. Therefore, the small signal linear model results:
$$\begin{cases} M\ddot{x}(t) + B\dot{x}(t) + Kx(t) - \frac{q_0}{\epsilon A}q(t) = f(t) \\ L\ddot{q}(t) + R\dot{q}(t) + \frac{1}{C_0}q(t) - \frac{q_0}{\epsilon A}x(t) = 0 \end{cases}$$
with $C_0 = \epsilon \frac{A}{x_0}$. Draw the bond graph and the block diagram of the linearized model.

**Exercise 7.7.** Figure 7.20 illustrates a hydraulic-mechanical system, where the mechanical subsystem comprises the elements mass, spring and damper represented in the figure by the symbols $M$, $K$ and $B$, respectively. This subsystem is subject to an externally imposed linear displacement $x_0$ that acts on the hydraulic part of the system with force $f$ by means of a piston with area $A$. Let the symbols $C$ and $R$ represent the capacitance of the tank and the hydraulic resistance, respectively. Additionally, the symbols $p_1$ and $p_2$ stand for pressures and $q$ is the volume flow rate. Moreover, the corresponding bond graph is shown in Figure 7.21 where the symbols $\alpha$, $\beta$, $\gamma$, $\delta$ and $\varepsilon$ denote elements of the system.

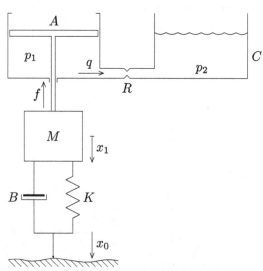

Figure 7.20 The hydraulic-mechanical system of Exercise 7.7.

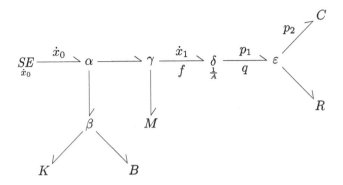

Figure 7.21 The bond graph of Exercise 7.7.

1. For the bond graph it comes that:

   **A)** $\alpha = 1$, $\beta = 0$, $\gamma = 0$, $\delta = TF$, $\varepsilon = 0$
   **B)** $\alpha = 0$, $\beta = 1$, $\gamma = 1$, $\delta = GY$, $\varepsilon = 1$
   **C)** $\alpha = 1$, $\beta = 0$, $\gamma = 0$, $\delta = GY$, $\varepsilon = 1$
   **D)** $\alpha = 0$, $\beta = 1$, $\gamma = 1$, $\delta = TF$, $\varepsilon = 0$

2. Assign the causality to the bond graph.

3. Find an analogous electrical circuit of the hydraulic-mechanical system in Figure 7.20.

**Exercise 7.8.** Consider the hydraulic-mechanical system shown in Figure 7.22, where the double rod hydraulic cylinder is connected to a two pulley system in order to act on mass $M_2$. An external hydraulic circuit provides the volume flows rates $q_1$ and $q_2$ that enable the cylinder to produce force $f_1$ and the linear displacement $x_1$. Let $C_i$ ($i = 1, 2$) and $p_i$ denote the cylinder's chambers capacitances and pressures, respectively, $A$ represent the cylinder piston area and $M_1$ the rod mass. The pulleys transform the translation motion provided by the cylinder, e.g., for the pulley on the right consider the relations $f_1 = n f_{12}$ and $x_1 = n x_{12}$, where $n$ is the transformation ratio. Moreover, the symbols $B$ and $K$ denote the viscous friction coefficient and the springs stiffness constant (Hooke's law), respectively.

Complete the given partial bond graph in Figure 7.23 and analyse the causality.

**158** ■ An Introduction to Bond Graph Modeling with Applications

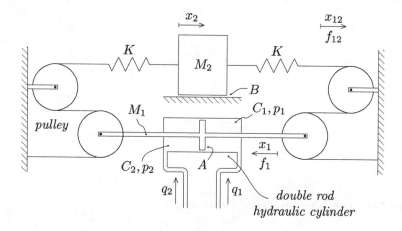

Figure 7.22 The hydraulic-mechanical system of Exercise 7.8.

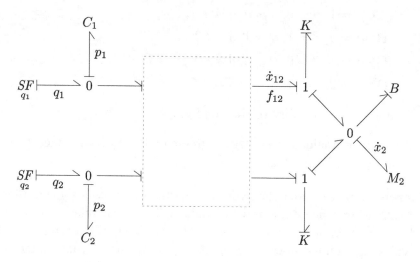

Figure 7.23 The partial bond graph of Exercise 7.8.

**Exercise 7.9.** The multi-domain system seen in Figure 7.24 uses three hydraulic cylinders that serve as the interface to the mechanical part of the system. The top cylinder is subject to the external force $f_1$ that produces the displacement $x_1$, while the two bottom cylinders act on the springs $K_2$ and $K_3$ with forces $f_2$ and $f_3$, respectively. Consider also that the piston of the middle cylinder has a fluid leakage modelled by the hydraulic resistance $R_2$. Moreover, the hydraulic circuit that connects to the bottom cylinder employs an ideal fluid transformer with the transformation ratio $\frac{A_4}{A_3}$, where $A_i$ ($i = 3, 4$) denotes the pistons areas. The remaining symbols include variables and elements with the same meaning as the ones described, and the additional symbols $M$, $B_j$ ($j = 1, 2, 3$), $p_j$ and $q_j$ that stand for the mass, viscous friction coefficients, pressures and volume flow rates, respectively. Find the bond graph of the system and apply the causal rules to assign the stokes to the bonds.

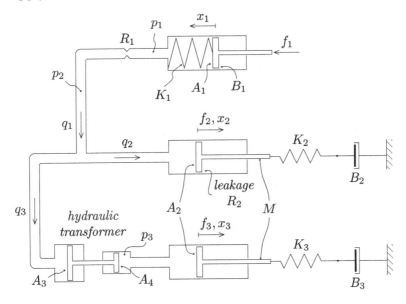

Figure 7.24 The multi-domain system of Exercise 7.9.

**Exercise 7.10.** The bond graph depicted in Figure 7.25 models an hydraulic-mechanical system driven by the pressure source $p_1$, that shows a causality conflict in the 0 junction. Let $R$ and $B$ stand for dissipator elements, while $L$ and $K$ represent effort stores and $M$ is a flow store. Moreover, the symbols $GY$ and $TF$ are two-ports elements with the transformation ratios $n_1$ and $n_2$, respectively. Find a system that can be represented by the bond graph and having solved the causality conflict. Consider that the final system remains linear for small displacements.

**Exercise 7.11.** Figure 7.26 illustrates a system that covers three energy domains. The external force $f$ acts on the piston (on the left) with mass $M_1$ and area $A_1$ that interacts with the hydraulic subsystem following the relations $f_1 = A_1 p_1$ and $q_1 = A_1 \dot{x}_1$. Let $f_1$, $\dot{x}_1$, $p_1$ and $q_1$ stand for force, linear velocity, pressure and volume flow rate, respectively. Consider that the piston with area $A_2$ (on the right) has the same dynamic relations as the piston on the left. Additionally, on the hydraulic part of the system

**160** ■ An Introduction to Bond Graph Modeling with Applications

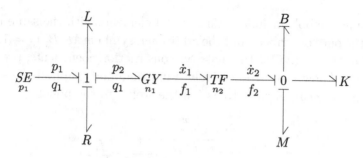

**Figure 7.25** The bond graph of Exercise 7.10.

$C_H$ and $R_H$ denote the capacitance of a tank and a hydraulic resistance, respectively. In the electrical part of the system an induced electromotive force $e = K_e \dot{x}_2$ emerges as the ferromagnetic core with mass $M_2$ moves, while on the mechanical part a force $f_3 = K_e i$ is generated as the current goes trough the coil. Moreover, in the electrical circuit $R$ is a resistance and $L$ denotes the inductance.

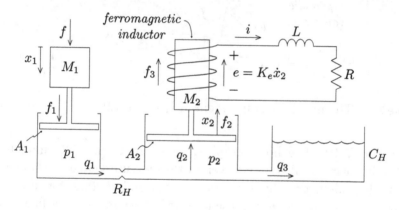

**Figure 7.26** The multi-domain system of Exercise 7.11.

1. Draw the corresponding bond graph and discuss the causality.

2. Determine the mathematical model and sketch the block diagram of the system.

**Exercise 7.12.** Figure 7.27 illustrates a hydroelectric power plant where the dike has a breakwater feature based on a porous wall. When filled with air, the alveolus help to ease the waves impact force on the wall. The electricity generating process starts from the fluid flow $q_0$ going through the pipe with hydraulic resistance $R_0$ and inertance $L_0$. The fluid then propels the turbine (under pressure $p_0$) that produces the torque $T$ and angular velocity $\omega$, that in turn, drives the electrical generator with resistance $R$ and inductance $L$. The induced electromotive force $e$ creates a current $i$ that supplies the electrical load $R_L$. Consider the relations $q_0 = K_T \omega$ and $T = K_T p_0$ (where $K_T$ is a constant) for the hydraulic to mechanical power transformation, and $\frac{1}{K_b}$ (where $K_b$ is a constant) as the ratio for the mechanical to electrical power variables transformation. Let the symbols $C_i$ ($i = 1, 2$), $R_i$, $L_i$ and $q_i$, represent the capacitances of the alveolus, hydraulic resistances, fluid inertances and flows, respectively. Draw the bond graph of the system considering the causality of the elements.

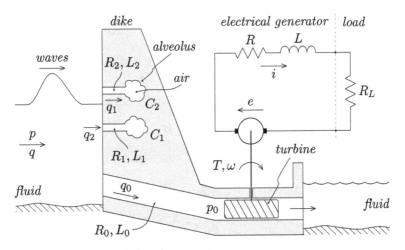

Figure 7.27  The hydroelectric power plant of Exercise 7.12.

**Exercise 7.13.** Consider the schematic diagram of the hydraulic-mechanical system shown in Figure 7.28. Let the symbols $A_i$ ($i = 1, 2, 3, 4$), $R_i$, $h_i$ and $q_i$ represent the areas of the tanks, hydraulic resistances, fluid heights and flow rates, respectively. Additionally, the symbol $L_j$ ($j = 1, 2$) stands for hydraulic inertances. The two upper tanks receive fluid from the external flow sources $q_{i_1}$ and $q_{i_2}$. Thereafter, the output flows ($q_1$ and $q_2$) feed another set of two tanks where the accumulated fluid drives a mechanical angle measuring device. The linear force produced by the fluid weight in each tank is given by $f = g\mu Ah$, where the symbols $g$ and $\mu$ denote the acceleration of gravity and the density of the fluid, respectively.

In the measuring device the symbols $l_1$ and $l_2$ are the length of the lever arms, $K_1$ and $K_2$ represent stiffness constants, $B$, $B_1$ and $B_2$ denote viscous friction coefficients and $J$ is a inertia. Moreover, the variables $f_j$, $T$, $x_j$ and $\theta$, denote force, torque, linear and angular displacement, respectively. Consider that the system remains linear for small angular displacements.

Find the bond graph of the system and discuss the causal relations between the elements.

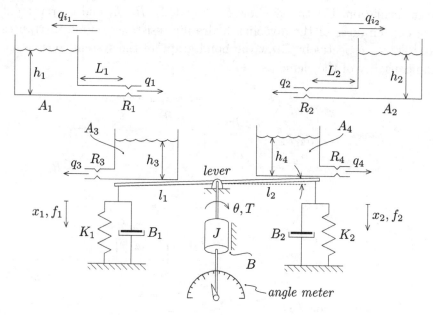

Figure 7.28 The hydraulic-mechanical system of Exercise 7.13.

**Exercise 7.14.** Consider the system shown in Figure 7.29 that represents a simplified nuclear power plant. To generate electricity the thermal process started by the nuclear reactor is used to create a water steam flow that drives a turbine. The output rotational mechanical power is then converted into electricity by means of an electrical generator. After leaving the turbine, the steam is cooled down in a condenser apparatus so the water can be reused in the heat exchange process.

The temperature in the core $\theta_0$ leads to a water stream producing temperature $\theta_1$ that occurs in the steam generator, where the height of the liquid water is kept constant by means of a controlled pump. Let the additional thermal process symbols $C_1$, $R_i$ ($i = 0, 1$) and $q_i$, stand for thermal capacitance, thermal resistances and heat flows, respectively.

The hydraulic gas produced in the steam generator is given by the relation $q_2 = \alpha\theta_1$, where $\alpha$ is a constant and $q_2$ is the fluid flow that is conducted to the turbine through the pipes with resistance $R_2$ and inertance $L_2$. Moreover, the turbine has a hydraulic resistance $R_3$ and sends the low pressure steam to a condenser. Here, the gas is subject to a controlled temperature so it can condense back to water in order to be pumped into the steam producing process. To maintain a constant water temperature, the condenser uses a controlled pump that supplies a variable cool water flow from an external source (e.g., a river).

The steam turbine produces a torque $T$ and angular velocity $\omega$ that drives the electrical generator through a rigid shaft with inertia $J$ and a friction $B$. Consider the transformation ratios $n_1$ and $n_2$ for the turbine and generator, respectively. The induced electromotive force $e$ creates a current $i$ in the electrical circuit with resistance $R$, inductance $L$ and the electrical transformer that has a $n_3$ ratio. The electricity distribution grid is represented by the load $R_L$.

Draw and assign the causality to the bond graph that models the electricity production process. Do not consider the power plant sub-processes that use the controlled pumps.

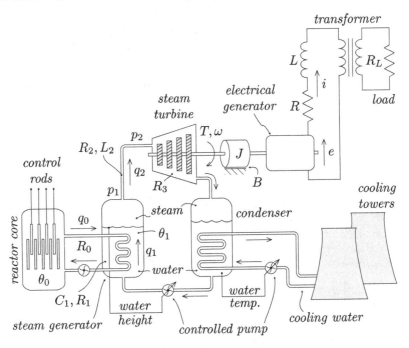

Figure 7.29 The nuclear power plant of Exercise 7.14.

**Exercise 7.15.** The multi-domain system shown in Figure 7.30 has two closed-loop controlled processes. The control action is done by means of proportional-integral-derivative (PID) controllers that output the calculated feedback error as $u(t) = K_p e(t) + K_i \int_0^t e(\tau) d\tau + K_d \frac{de(t)}{dt}$, where $K_p$, $K_i$ and $K_d$, denote the coefficients for the proportional, integral, and derivative terms, respectively. The PID on the left controls the voltage applied to the DC motor armature given two pressure inputs, the reference value $p_{ref}$ and the current pressure $p_2$ in tank 2. The PID on the right side of the figure is responsible for acting on the temperature feedback error $(\theta_{ref} - \theta_2)$ in order to control the voltage applied to the heater device in tank 1.

The PID controlled DC motor (with inductance $L_m$, resistance $R_m$ and back-emf $e_b = K_b \omega_m$) drives the fixed displacement hydraulic pump through a rigid shaft with torque $T_m = K_i i_m$ and angular velocity $\omega_m$, where $K_b$ is a DC motor constant and $K_i = K_b$. The transformation from rotational movement to hydraulic power is given by the relations $p_i = \frac{1}{K_h} T_m$ and $q_i = K_h \omega_m$, where $K_h$ is a constant related to the pump's capacity and the symbols $p_i$ and $q_i$ represent the output pressure and flow rate of the pump, respectively. Let the additional symbols $C_i$ ($i = 1, 2$), $R_i$, and $q_i$, represent tank capacitances, hydraulic resistances and fluid flow rates, respectively. Related to the temperature controlling process the symbol $C_T$ stands for thermal capacitance, $R_T$ is a thermal resistance and $q_T$ represents the heat flow that happens from the $\theta_1$ and $\theta_2$ temperatures difference. Find the bond graph and assign the causality.

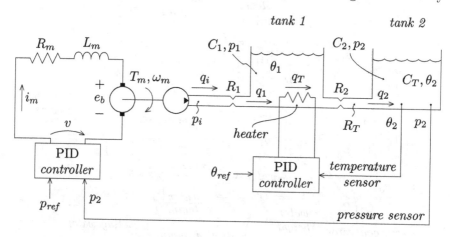

Figure 7.30  The multi-domain system of Exercise 7.15.

**Exercise 7.16.** Figure 7.31 depicts an hydraulic-mechanical system that employs two PID controllers that consist of a proportional, integral and derivative gains. This type of closed-loop control uses two input signals to produce the output $u$ (in the time domain) that is fed back to the process as the controlling signal. In sum, the difference $e(t)$ between the input variables is processed accordingly to the equation $u(t) = K_p e(t) + K_i \int_0^t e(\tau) d\tau + K_d \frac{de(t)}{dt}$, where $K_p$, $K_i$ and $K_d$, denote the coefficients for the proportional, integral, and derivative terms, respectively. In this system the PID located at the top controls the aperture of the hydraulic resistance $R_2$ based on the desired pressure value $p_{ref}$ and the actual hydraulic actuator chamber pressure $p_2$. On the other hand, the bottom controller takes as inputs the reference $\dot{x}_{ref}$ value for the velocity of mass $M_3$ and the value $\dot{x}_3$ obtained from the linear velocity sensor. The controller then outputs a signal that controls the hydraulic pump, i.e., the fluid flow $q_i$ that drives the hydraulic part of the system.

Consider for the hydraulic subsystem the symbols $C_i$ ($i = 1, 2$), $R_i$, $p_i$ and $q_i$, that denote capacitances, hydraulic resistances, pressures and fluid flow rates, respectively. In the translational mechanical part of the system $f_1$ and $f_2$ are forces, $x_1$, $x_2$ and $x_3$ denote displacements, $K_1$, $K_2$ and $K_3$ stand for spring stiffness constants (Hooke's law) and, $B_1$ and $B_2$ are viscous friction coefficients. Additionally, the rigid bodies masses are represented by the symbol $M_i$ ($i = 1, 2, 3$), while $A$ is the piston area of the hydraulic actuator. Find the bond graph and study the causality.

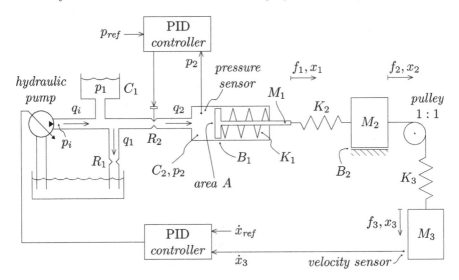

Figure 7.31   The hydraulic-mechanical system of Exercise 7.16.

**Exercise 7.17.** Consider the following questions where a representation of the human body is analyzed under several energy domains.

1. Figure 7.32 depicts the human body as a mechanical system, that has been divided into six sections: (1) head and neck, (2) trunk, (3) arms, (4) hands, (5) legs and (6) feet. Each of the considered sections is modeled by a mass and joints, comprising elasticity and friction elements in parallel, that connect it to the adjacent body section(s). For simplicity consider that both arms, legs, feet and hands are each a single body section.

   Let the symbols $B$, $K$ and $M$ stand for the viscous friction coefficients, the stiffness constants (Hooke's law) and the masses, respectively. The body has five friction elements $\{B_{12}, B_{23}, B_{25}, B_{34}, B_{56}\}$, five elasticity elements $\{K_{12}, K_{23}, K_{25}, K_{34}, K_{56}\}$ and six masses $\{M_1, M_2, M_3, M_4, M_5, M_6\}$. Additionally, consider that

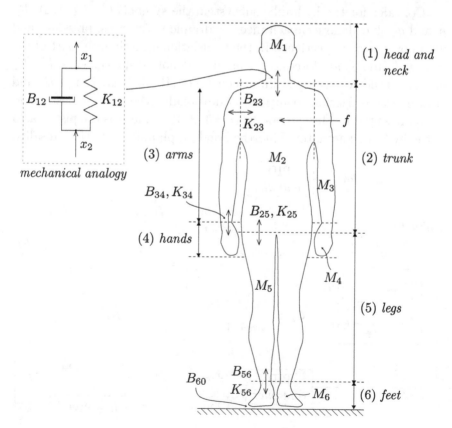

Figure 7.32 The mechanical system of Exercise 7.17.

between the feet and the floor there is a friction $B_{60}$, and that an external force $f$ is applied to the trunk. Find the bond graph and assign the causality.

2. Adopting the {velocity, force}→{voltage, current} analogy, sketch an analogous electrical circuit of the human body as a mechanical system given in Figure 7.32.

3. Consider the illustration of the human body as a thermal system shown in Figure 7.33. For this analysis consider seven sections: surrounding ambient, head and neck, trunk, arms, hands, legs and feet. Each section is represented by the index 0 to 6, respectively, and for simplicity consider that both arms, legs, feet and hands are each a single body section. Let the symbols $\theta$, $R$ and $C$ stand for the temperature, the thermal resistance and the thermal capacitance, respectively. The system has seven temperatures $\{\theta_0, \theta_1,$

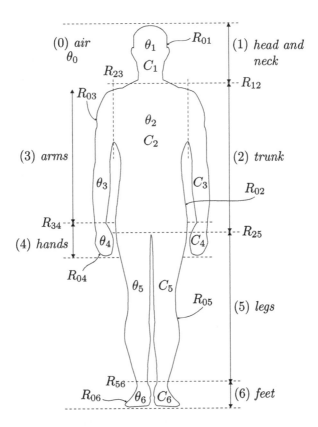

Figure 7.33 The thermal system of Exercise 7.17.

$\theta_2, \theta_3, \theta_4, \theta_5, \theta_6\}$, eleven thermal resistances $\{R_{01}, R_{02}, R_{03}, R_{04}, R_{05}, R_{06}, R_{12}, R_{23}, R_{25}, R_{34}, R_{56}\}$ and six thermal capacitances $\{C_1, C_2, C_3, C_4, C_5, C_6\}$. Furthermore, consider that the ambient temperature $\theta_0$ is constant. Find the bond graph of the thermal system and assign the causality.

4. Sketch an analogous electrical circuit of the thermal system in Figure 7.33 using the {temperature, heat rate}→{effort, flow} analogy.

**Exercise 7.18.** Consider the next set of questions where a simplified plant leaf is analyzed under several energy domains. For this exercise the leaf's veins where divided into five sections, numbered from 1 to 5 in the questions figures. Furthermore, consider that the blade has neglectable characteristics.

1. Figure 7.34 depicts the plant leaf as a linear mechanical system. For this analysis consider the mechanical analogy shown in the figure used to model the first section of the leaf's veins. Moreover, let $K_j$ ($j = 1, 2, 3, 4, 5$) be the stiffness constants (Hooke's law), $B_j$

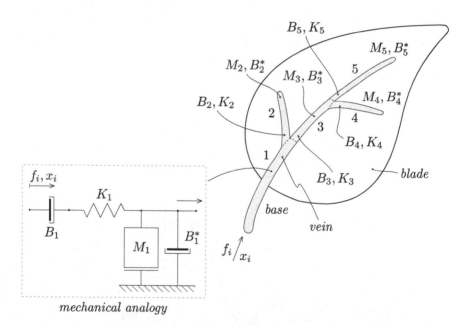

Figure 7.34 The plant leaf analyzed as a mechanical system of Exercise 7.18.

and $B_j^*$ the viscous friction coefficients and $M_j$ the linear inertias. The base of the leaf is subject to the external force $f_i$, causing the linear displacement $x_i$. Find the bond graph and discuss the causality.

2. Consider the drawing of the plant leaf as a hydraulic system shown in Figure 7.35. In the figure there is also represented the hydraulic analogy used to model the first section of the leaf's veins. Additionally, let $L_j$ stand for the inertances, $R_j$ ($j = 1, 2, 3, 4, 5$) and $R_j^*$ for the hydraulic resistances and $C_j$ for the capacitances. At the base of the leaf there is a constant pressure $p_i$ and the flow rate $q_i$. Find the bond graph of the hydraulic system and assign the causal strokes.

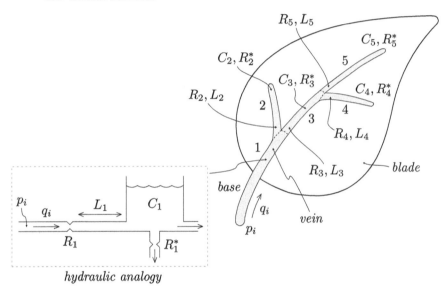

Figure 7.35 The hydraulic system of Exercise 7.18.

3. Draw an analogous electrical circuit of the hydraulic system in Figure 7.35. Use the correspondence {pressure, flow rate}→{voltage, current}.

CHAPTER 8

# Bond Graph Modeling and Simulation Using 20-sim

## 8.1 INTRODUCTION

This chapter offers a bond graph guided example on how to model and study the behavior of a physical system using the 20-sim® software application (version 4.8) [32], available at www.20sim.com. However, for the same purpose other options exist [33, 34, 35, 36, 37, 38]. The 20-sim software is a modeling and simulation tool used to analyse dynamical systems from multiple physical domains, including systems that are a combination of several. The systems models can be constructed graphically, e.g., in the form of bond graphs. A library of standard components is available that can be added and combined into a model in order to represent the dynamics of the power variables in the system. The provided linear bond graph elements can be edited, and new elements can also be created.

Note that the 20-sim software adopts the 'Option 2' analogy {force/torque, linear/angular velocity}→{effort, flow} for translational and rotational mechanical systems. Appendix B provides correspondence tables for modeling systems using the 'Option 1' analogy with this software application.

Note: Take into account that you will not be able to save your work if you are using the demonstration license of 20-sim.

## 8.2 GUIDED EXERCISE

Consider the electromechanical system shown in Figure 8.1 that comprises an armature controlled DC motor with the inertia $J_m$ and rotational friction $B_m$. Coupled to the motor's axis we have one gear that drives a load with the inertia $J_L$ and rotational friction $B_L$. The load axis is not rigid and has some elasticity represented by the spring with stiffness constant $K$. Moreover, the load requires an additional torque $T_L$. The gear has a transformation ratio $n = \frac{N_1}{N_2}$, where $N_i$ ($i = 1, 2$) is the number of teeth of the $i$-th gear-wheel. Consider that the DC motor provides the torque $T_m = K_T i_a$ and has a back electromotive voltage $e_b = K_b \dot{\theta}_m$ (consider $K_T = K_b$) for the angular displacement and velocity $\theta_m$ and $\dot{\theta}_m$, respectively.

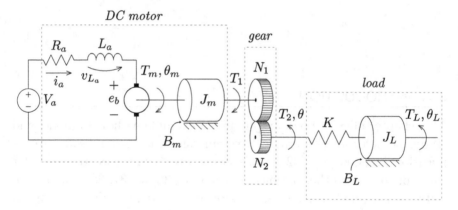

Figure 8.1  The diagram of the electromechanical system.

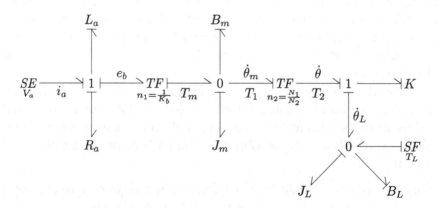

Figure 8.2  The bond graph of the electromechanical system.

The bond graph that models this system is shown in Figure 8.2 and the alternative classical mathematical model representation of the system is as follows:

$$\begin{cases} V_a = R_a i_a + v_{L_a} + e_b \\ i_a = \frac{1}{L_a} \int_0^t v_{L_a}(\tau) \, d\tau \\ i_a = n_1 T_m, \ \dot{\theta}_m = n_1 e_b, \ n_1 = \frac{1}{K_b}, \text{ with } K_b = K_T \\ T_m = J_m \ddot{\theta}_m + B_m \dot{\theta}_m + T_1 \\ T_1 = n_2 T_2, \ \theta = n_2 \theta_m, \ n_2 = \frac{N_1}{N_2} \\ T_2 = K(\theta - \theta_L) \\ T_2 = J_L \ddot{\theta}_L + B_L \dot{\theta}_L + T_L \end{cases}$$

The alternative classical block diagram representation of the system is shown in Figure 8.3, where $V_a(s) = \mathscr{L}[V_a(t)]$ and $T_L(s) = \mathscr{L}[T_L(t)]$ are the Laplace transforms of the input signals and $\Theta_L(s) = \mathscr{L}[\theta_L(t)]$ is the Laplace transform of the output signal.

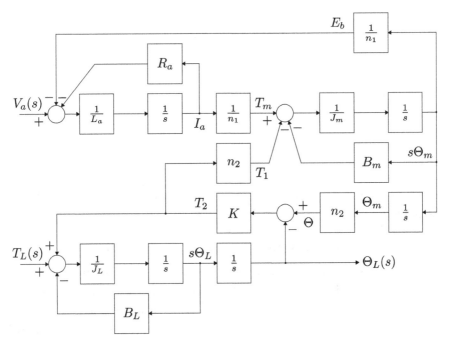

Figure 8.3 Block diagram of the electromechanical system.

## System implementation in the 20-sim software application

1. Build the bond graph model shown in Figure 8.2 using the 20-sim software application (version 4.8) [32, 39]. This software, among other features, allows the modeling and simulation of dynamical systems using the bond graph approach.

    (a) Start the 20-sim software application.

    (b) The window that opens up (Figure 8.4) is called the **20-sim Editor** and is used to enter and edit the models.

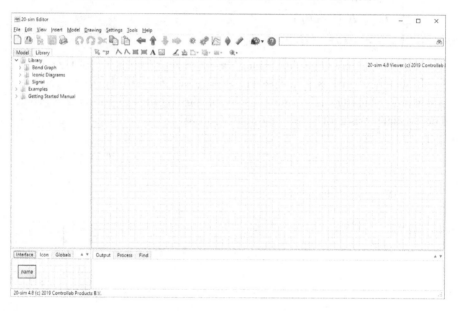

Figure 8.4  The 20-sim software application main window, the **20-sim Editor**.

    (c) The **20-sim Editor** window consists of four parts:
    
    - **Model/Library** tabs – located on the left area of the window, the **Model** tab shows the current model as an hierarchical tree. The **Library** tab shows the available 20-sim library elements grouped in 3 folders (**Bond Graph**, **Iconic Diagrams** and **Signal**).
    - Graphical/Equation Editor – the main checkered area of the **20-sim Editor** window is called the Graphical Editor and is used to enter and edit the graphical parts of your models. When a lower level element of the model

hierarchy is selected this area will show the model equations (Equation Editor).

- **Output/Process/Find** tabs – located on the bottom area of the main window, the **Output** tab shows the files that are opened and stored. The **Process** tab shows the compiler messages and the **Find** tab shows the search results.
- **Interface/Icon/Globals** tabs – on the lower left corner of the main window the **Interface** tab shows the interface (inputs, outputs, ports) of a selected element. The **Icon** tab shows the icon of a selected element and the **Globals** tab will show the global parameters and variables of your model.

(d) To build the system bond graph model, you need to use the main checkered area in the **20-sim Editor** window (Graphical Editor) to place the bond graph elements that are available in the **Library** tab.

- Alternatively you can also access the elements library by selecting the **Open Browser** option located in the **File** menu.

(e) Start building the bond graph by expanding the **Bond Graph** folder on the **Library** tab to expose all the available bond graph elements that can be dragged and dropped into the Graphical Editor. To insert elements you can also use the **Insert** option available in the main window menu or by right clicking on the checkered area.

(f) As an example, to add a bond graph junction element to the model right-click on the checkered area and select **Insert**. Follow with the selection of the **Knot** option and then the **One Junction** suboption.

(g) This will add a **1** junction multi-port element to the **20-sim Editor** window as shown in Figure 8.5.

(h) Proceed by inserting the remaining bond graph elements needed to build the model shown in Figure 8.2 (refer to Appendix B). The 20-sim bond graph library has:

- Junctions (**0** junction, **1** junction): elements that couple energy between various other elements.

**176** ■ An Introduction to Bond Graph Modeling with Applications

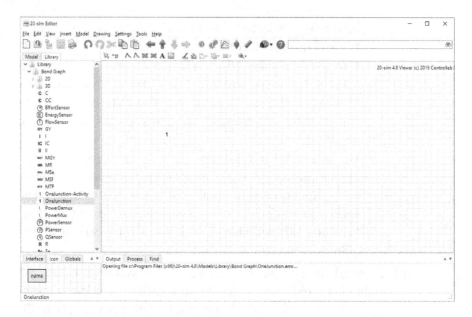

Figure 8.5 The **20-sim Editor** window after adding a 1 junction.

- Stores (**C**, **I**): elements that store energy. Examples of **C** elements are a mechanical inertia and an electrical capacitor, and examples of **I** elements are a mechanical spring and an electrical inductance.
- Dissipator (**R**): elements that dissipate energy.
- Sources (**Se**, **Sf**): elements that generate energy.
- Transformer (**TF**): elements that convert energy.

(i) From the **Bond Graph** folder of the **Library** tab, select the desired elements and drag them into the Graphical Editor, arranging them in the way shown in Figure 8.6.

(j) Enter a name for each of the bond graph elements. To do this, right-click on the desired element and in the context menu that appears select the **Properties** option. In the *Name* textbox, of the **Description** tab, enter the name of the element as defined in the bond graph of Figure 8.2.

(k) Your model should look like the one shown in Figure 8.7.

(l) Establish the energy bonds between the elements and the junctions:

- Select the **Connect** icon located in the Graphical

Bond Graph Modeling and Simulation Using 20-sim ■ 177

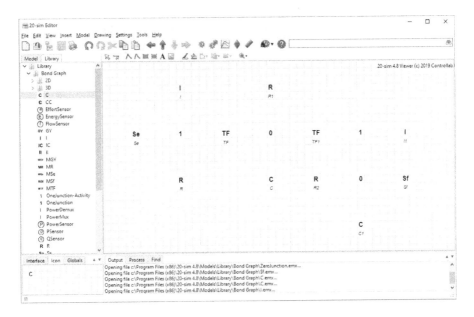

**Figure 8.6** The elements needed to implement the model of the electromechanical system.

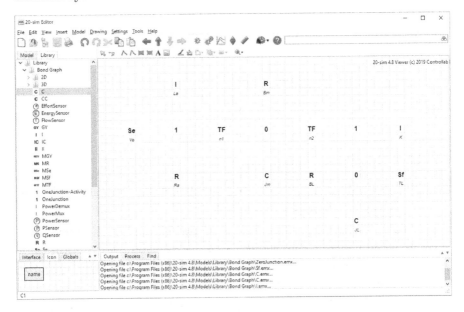

**Figure 8.7** The elements needed to implement the model of the electromechanical system, with names assigned.

Editor window toolbar to change to connect mode. Alternatively, you can use the keyboard spacebar to switch between the Select and Connect modes.

- Left-click the effort source **Se** that represents the supply voltage $V_a$, and then left-click the **1** junction element located to the right.
- An energy bond that connects both elements should be visible as shown in Figure 8.8.

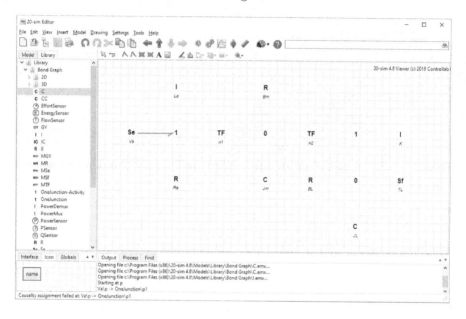

Figure 8.8 Assignment of the energy bond between the effort source **Se** and the **1** junction.

(m) Establish all the necessary bonds until your model looks like the one shown in Figure 8.9.

(n) In order to prevent the visual overlapping of the bond with the name of the element you can use the **Show Name** option, available by right-clicking on the element, to change the position of the name.

(o) Causality is automatically assigned by the 20-sim software. However, it can be changed by the user if needed. In the case of mechanical systems we find the authors assigning the flow and effort variables to the force/torque and velocity, respectively, or vice-versa. In our case we adopt flow for force/torque

# Bond Graph Modeling and Simulation Using 20-sim ■ 179

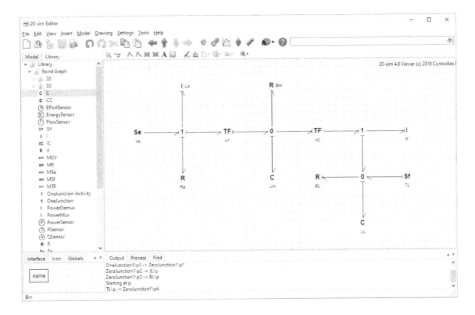

Figure 8.9 Electromechanical system model with all the bonds established and causality assigned.

and effort for velocity, that is the opposite of the 20-sim default settings. Therefore, implement the following action if necessary:

- Go into Select mode by using the icon.
- Right-click the bond you want to change the causality. For example, select the bond that connects to the **R** element (resistance $R_a$) and then choose the option **Properties**.
- In the window that appears (Figure 8.10) select the type of causality you want.

(p) After finishing the bond graph you must verify its integrity:

- Select the option **Check Complete Model** from the **Model** menu or use the icon located in the main window toolbar.
- The 20-sim then checks the model automatically. If an error is detected, its description is displayed in the **Process** tab (bottom area of the main window).

Figure 8.10 **Bond Properties Editor** window.

2. Check the equations of the model. They define the relationship between the effort and flow variables for each bond graph element.

   (a) Select the **Model** tab to shown the hierarchical tree of the model. Left-click the element for which you want to analyse the equations. For example, select the **R** element that stands for the DC motor mechanical friction $B_m$.

   (b) The Graphical Editor switches to the Equation Editor view and displays the equations of the selected element, as shown in Figure 8.11. Consider that the terms p.e. and p.f. stand for 'power effort' and 'power flow', respectively.

   (c) Review each one of the elements and correct if necessary.

   (d) To return to the Graphical Editor view select **model** at the top of the hierarchical tree.

## System analysis using the 20-sim software application

1. The next task is the simulation and analysis of the time domain response of the system to a step input signal. For this study consider the system parameters values listed in Table 8.1.

   (a) Select the **Show Parameters** option from the **Model** menu or use the icon ◆ located in the main window toolbar to set the initial values of the system parameters.

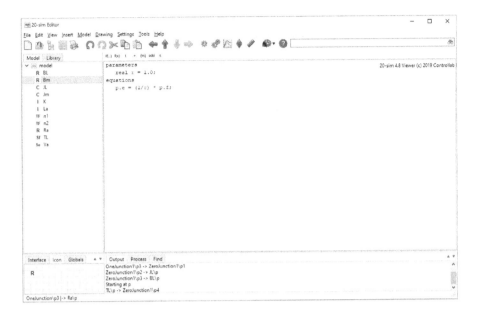

Figure 8.11 Window showing the equations of the selected **R** element ($B_m$).

TABLE 8.1 Initial numerical values for the parameters of the electromechanical system in Figure 8.1.

| Element | Value | Units |
|---------|-------|-------|
| $V_a$   | 12.0  | V |
| $L_a$   | 0.8   | H |
| $R_a$   | 8.0   | $\Omega$ |
| $n_1$   | 0.5   | |
| $B_m$   | 0.5   | $N \cdot m \cdot s/rad$ |
| $J_m$   | 1.0   | $kg \cdot m^2$ |
| $n_2$   | 0.9   | |
| $K$     | 1.0   | $N \cdot m/rad$ |
| $J_L$   | 0.4   | $kg \cdot m^2$ |
| $B_L$   | 0.8   | $N \cdot m \cdot s/rad$ |
| $T_L$   | 0.0   | $N \cdot m$ |

(b) Enter the numerical values for each element, such as depicted in Figure 8.12.

(c) After editing the values in the **Value** column, the window can be closed by doing **OK**.

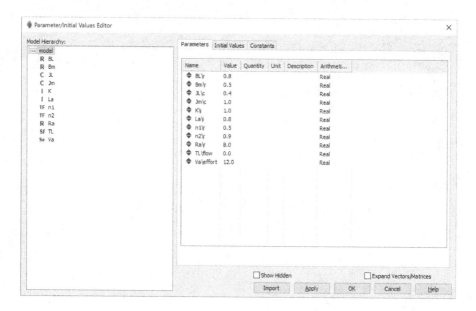

Figure 8.12 **Initial Values Editor** window of the system with the parameters numerical values.

(d) To set the simulation and plot the parameters follow the steps:

- Select the **Simulator** option from the **Tools** menu or, alternatively, select the icon located in the main window toolbar. The **20-sim Simulator** window will appear, like the one shown in Figure 8.13.
- In the **20-sim Simulator** window select the **Run** option from the **Properties** menu or use the icon of the window's toolbar to set the simulation parameters.
- Change the default values to those shown in Figure 8.14.
- To set the plots parameters select the **Plot** option from the **Properties** menu or use the icon.
- Select the **Plot Properties** tab and change the values to those shown in Figure 8.15.
- Select the **Y-Axis** tab and then click the **Choose** button to open the **Variable Chooser** window, as shown in Figure 8.16. Here you can select the system variable that will be plotted.

## Bond Graph Modeling and Simulation Using 20-sim ■ 183

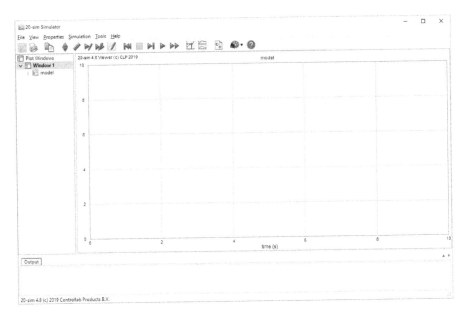

Figure 8.13   The **20-sim Simulator** window.

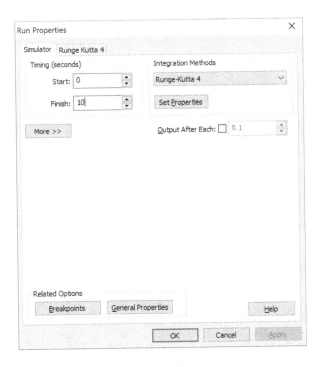

Figure 8.14   The 20-sim window to set the simulation parameters.

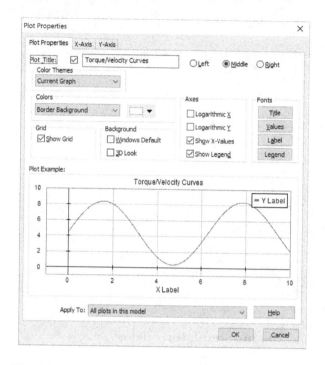

Figure 8.15  The 20-sim window to set the plot parameters.

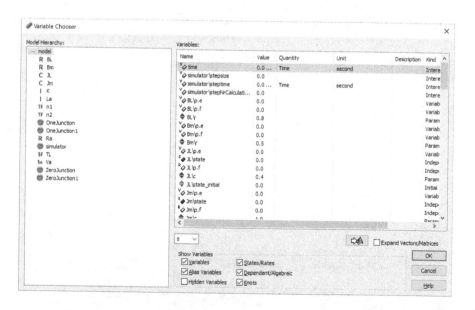

Figure 8.16  The 20-sim window for choosing the system variables to plot.

- From the list, select the variable *ZeroJunction1\effort* that corresponds to the angular velocity $\dot{\theta}_L$ of the load (elements $J_L$, $B_L$ and $T_L$). Click on the **OK** button to close the **Variable Chooser** window.
- To have a more informative plot legend fill the *Label* textbox, as seen in Figure 8.17.

Figure 8.17 **Plot Properties** window after adding the curve for the load angular velocity $\dot{\theta}_L$.

- Use the **Add Curve** button to repeat the procedure so that more system variables can be plotted in each simulation run. Related to the same 0 junction that connects the elements $J_L$, $B_L$ and $T_L$, add the curves for:
  i. The flow variable that stands for the motor torque delivered to the inertia $J_L$ (and friction $B_L$).
  ii. The flow variable that represents the inertia torque due to the element $J_L$.
  iii. The flow variable corresponding to the friction torque, that results from the action of the element $B_L$.

- From the energy bond that connects the element **TF** (with $n_2$) to the **1** junction, add a curve to plot the effort variable that represents the angular velocity $\dot{\theta}$.
- Click on the **OK** button to close the **Plot Properties** window.

(e) To run the simulation select the **Run** option from the **Simulation** menu or select the icon ▶ from the toolbar.

(f) Figure 8.18 shows the transient response of the selected variables: the load angular velocity $\dot{\theta}_L$, the load axis angular velocity $\dot{\theta}$, the motor torque delivered to the load, the load friction torque and the load inertia torque.

(g) Discuss the results of the simulation.

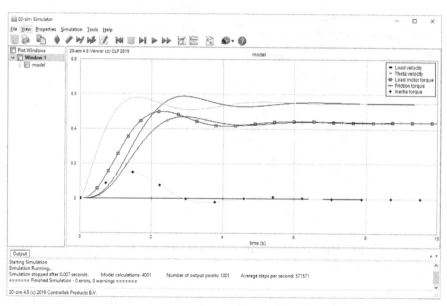

Figure 8.18 Simulation results showing the time evolution of the selected effort and flow variables for a step input signal $V_a$.

2. Perform the simulations needed to analyse the behavior of the load velocity when (i) the supply voltage is changed, (ii) the moment of inertia of the load is changed, (iii) the friction coefficient is changed, and (iv) the externally applied torque changes.

(a) Take note of the values of the steady-state load angular velocity $\dot{\theta}_L$ for the supply voltage values in Table 8.2. For the remaining bond graph elements use the initial values listed in Table 8.1.

- To change the parameters values for each simulation select the **Parameters** icon located on the toolbar of the **20-sim Simulator** window, and then use the **Apply** button.
- Use the **Numerical Values** option from the **View** menu to get the values of the angular velocity $\dot{\theta}_L$. Alternately, you can use the icon to open the **Numerical Values** window.

**TABLE 8.2** Values of the supply voltage $V_a$ to adopt in the study of the load angular velocity $\dot{\theta}_L$ (use the 'Angular velocity' column to write down the numerical values).

| Voltage source $V_a$ (V) | Angular velocity $\dot{\theta}_L$ (rad/s) |
|---|---|
| 1 | |
| 5 | |
| 10 | |

(b) Use the same procedure to study how the steady-state values of the load angular velocity $\dot{\theta}_L$ change for the different load inertia values in Table 8.3.

- Comment on the simulation results obtained for $J_L = 10$ kg · m².

**TABLE 8.3** Values of the inertia $J_L$ to use in the study of the load angular velocity $\dot{\theta}_L$ (use the $\dot{\theta}_L$ column to take note of the numerical values).

| Inertia $J_L$ (kg · m²) | Angular velocity $\dot{\theta}_L$ (rad/s) |
|---|---|
| 0.8 | |
| 2 | |
| 10 | |

(c) Write down the values of the steady-state load angular velocity $\dot{\theta}_L$ for the friction coefficient values in Table 8.4. Continue using the initial values listed in Table 8.1 for the remaining elements.

(d) Now to study how different values of the externally applied torque $T_L$ influence angular velocity of the load, use the $T_L$ values presented in Table 8.5.

TABLE 8.4  Values for the viscous friction coefficient $B_L$ to study on how they influence the load angular velocity $\dot{\theta}_L$ (use the 'Angular velocity' column to take note of the numerical values).

| Friction $B_L$ · (N · m · s/rad) | Angular velocity $\dot{\theta}_L$ (rad/s) |
|---|---|
| 1 | |
| 2 | |
| 5 | |

- Comment on the simulation results obtained for $T_L = -1.5$ N · m.

(e) Discuss the results of the simulations.

TABLE 8.5  Values for the torque $T_L$ to adopt in the study of the load angular velocity $\dot{\theta}_L$ (use the $\dot{\theta}_L$ column to write down the numerical values).

| Torque $T_L$ (N · m) | Angular velocity $\dot{\theta}_L$ (rad/s) |
|---|---|
| −0.2 | |
| −0.5 | |
| −1.5 | |

3. Now consider that the input voltage source $V_a$ changes from a step to a pulse type signal. Study the temporal evolution of the system for different durations of the input voltage for the case where $T_L$ represents a centrifugal load (e.g., a fan). This type of loads create an opposing torque that has a quadratic relation with the rotational velocity, $T_L = k\dot{\theta}_L^2$, where $k \in \mathbb{R}^+$ is a constant.

   (a) In the 20-sim software, to study a bond graph model with a time variant input signal, the standard (constant) energy sources must be replaced by modulated energy sources connected to a signal generator element.

   (b) From the **Bond Graph** folder of the **Library** tab, select the modulated source element **MSe** and **MSf** to replace, respectively, the **Se** and **Sf** sources present in your bond graph model.

   (c) In the **Library** tab open the **Signal** folder and then, from the **Sources** folder select the **SignalGenerator-Pulse** block and add it to the bond graph. Connect this block to the **MSe** element so it can provide a signal to modulate the voltage applied to the DC Motor armature.

(d) Your bond graph model should look like the one shown in Figure 8.19.

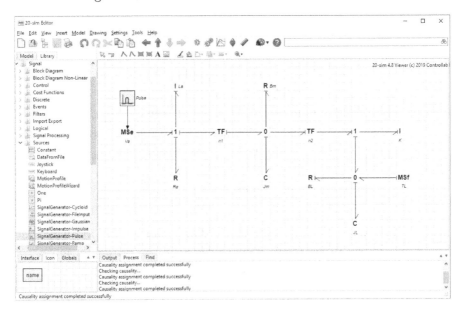

**Figure 8.19** The bond graph after replacing the sources with the modulated counterparts and a pulse signal generator.

(e) To generate the signal that must be provided to the **MSf** element in order to emulate the centrifugal load, do the following:

i. From the **Block Diagram Non-Linear** folder, located in the **Signal** folder, select the **Function-SquareSign** block and place it near the **MSf** element of the bond graph. This block outputs the square of the input signal, with sign and multiplied by a parameter $p$.

ii. In order to generate the signal $T_L$, the load angular velocity $\dot{\theta}_L$ must be obtained from the bond graph and used as the input for the **Function-SquareSign** block. The 20-sim Bond Graph library offers several elements that can provide effort and flow based signals to elements of the Signal library:

- **e** (EffortSensor): outputs the effort of a bond

**190** ■ An Introduction to Bond Graph Modeling with Applications

- **f** (FlowSensor): outputs the flow of a bond
- **p** (Psensor): outputs the integral of the effort of a bond
- **q** (Qsensor): outputs the integral of the flow of a bond
- **0** (ZeroJunction): outputs the junction common effort
- **1** (OneJunction): outputs the junction common flow

iii. Select the **EffortSensor** element from the **Bond Graph** folder and drop it exactly on the middle of the bond that connects the $J_L$ element to the **0** junction. Alternatively, the effort variable that represents $\dot{\theta}_L$ can be obtained directly from the **0** junction.

iv. Arrange and connect the elements as shown in Figure 8.20.

(f) Right-click the *Pulse* block and, from the context menu that shows up, select the **Parameters** option. Set the initial values as shown in Figure 8.21.

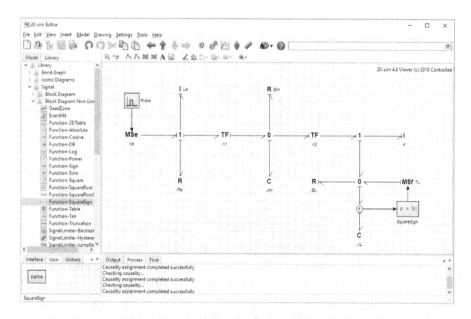

Figure 8.20 The bond graph with the signal generator elements connected to the modulated source elements.

Figure 8.21 Window to set the initial parameters values of the **SignalGenerator-Pulse** block.

(g) Repeat the procedure to set the initial values of the ***Square-Sign*** block as shown in Figure 8.22.

(h) Add the load torque $T_L$ to the signals that are plotted in each simulation run by performing the following steps:

   i. Use the icon located in the **20-sim Editor** window toolbar to open the **20-sim Simulator** window.
   
   ii. Select the **Y-Axis** tab of the **Plot Properties** window that shows up after selecting the plot icon.
   
   iii. Use the **Add Curve** button to add the signal for the flow variable of the bond that connects the **MSf** element to the **0** junction (i.e., the torque $T_L$).

(i) Select the **Run** option from the **Simulation** menu or select the icon from the toolbar to start the simulation. Figure 8.23 shows the output plot of the selected variables.

(j) Comment the results of the simulation.

(k) Analyse the waveforms of the monitored variables as the duration of the input voltage signal changes. For this study use the values 2 s, 5 s and 10 s for the 'stop_time' parameter of the ***Pulse*** block.

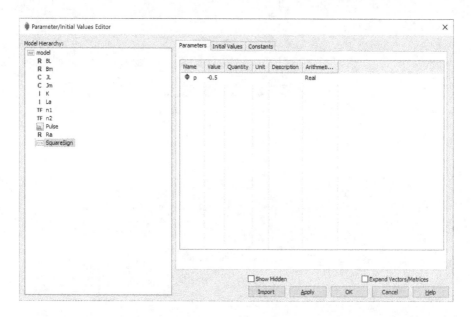

Figure 8.22 Window to set the initial parameters values of the **Function-SquareSign** block.

- To have a better view of the generated curves change the simulation duration to 20 s (recall Figure 8.14).
- Tip: use the reset simulation icon ◄◄ to clear the plot area before each new simulation.

(1) Discuss the results of the simulations focusing on the observed behavior when the input signal ceases.

4. Study the frequency response of the mechanical subsystem of the electromechanical system shown in Figure 8.1 using the 20-sim Frequency Domain Toolbox. In the 20-sim software, a symbolic linear-system can be derived by means of linearization out of an existing (non-linear) model. The resulting linear system can be studied using the Linear System Editor tool.

(a) In the **20-sim Simulator** window go to the **Tools** menu, then select the **Frequency Domain Toolbox** menu option. Select the **Model Linearization** suboption to open the **20-sim Model Linearization** window.

(b) The mechanical subsystem in Figure 8.1 comprises the DC

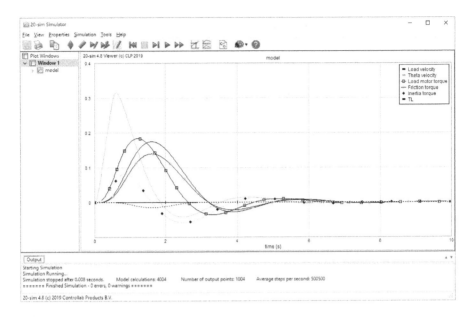

Figure 8.23 Simulation results showing the evolution of the selected effort and flow variables for a pulse type input signal $V_a$ and a centrifugal load torque (dashed line).

Motor axis, the gears and the load axis. To linearise this subsystem select as the input variable the motor torque $T_m$ and the load angular velocity $\dot{\theta}_L$ as the output variable.

(c) Figure 8.24 illustrates the settings needed to linearize the system to study. After clicking on the **OK** button, the **20-sim Linear System Editor** window (Figure 8.25) is presented.

(d) The Linear System Editor of the 20-sim software allows you to enter or edit linear time-invariant models. It supports continuous and discrete-time single input/single output (SISO) systems with and without time delay. Like in our case, the linear system can also be the result of a linearization operation in the 20-sim Simulator. It also offers several plot options to study the time and frequency responses of the linear system, namely: unit step response, Bode plot, Nyquist plot, Nichols chart and Pole/Zeros plot (including Root Locus).

(e) Select the **Bode** button (located on the **Plots** interface controls group) to compute the system's Bode diagrams. This

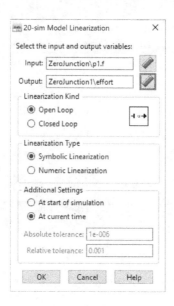

Figure 8.24 The **20-sim Model Linearization** window settings to linearize the system to study.

Figure 8.25 The **20-sim Linear System Editor** shows (by default) the Transfer Function of the linearized system.

plot shows the output amplitude and phase of the linear system as function of the input signal frequency.

(f) The 20-sim software shows a window like the one in Figure 8.26. In the **20-sim Bode Plot** window right-click on the plot area and selected the option **Phase/Margin Gain** option to display the gain margins of the system.

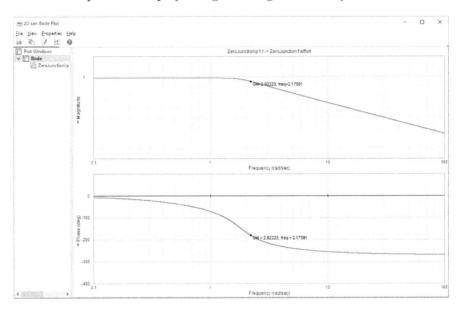

Figure 8.26 Result of the frequency domain study represented in the **20-sim Bode Plot** window.

(g) Analyse the frequency domain response of the system as the load axis spring stiffness constant $K$ changes. Use the following values for the $K$ parameter: 0.5 N · m/rad, 0.1 N · m/rad and 0.01 N · m/rad.

i. Select the parameters button (located on the **System Description** interface controls group) in order to access the linear system **Parameter/Initial Values Editor** window (Figure 8.27).

ii. After changing the value of the $K$ parameter, select the **OK** button to generate the new Bode plot.

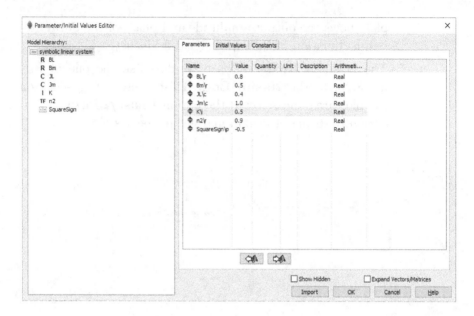

Figure 8.27 The **Parameter/Initial Values Editor** window used to edit the linearized system parameters values.

(h) Discuss the results of the simulations.

# Solutions

## CHAPTER 3 SOLUTIONS

**Exercise 3.1:** The correct answer is option **B)** since $\alpha = C_1$, $\beta = R_2$ and $\gamma = C_2$.

**Exercise 3.2:** The correct answer is option **D)** given that $\alpha = C_1$ and $\delta = R_1$. Figure S.1 shows the bond graph including all symbols.

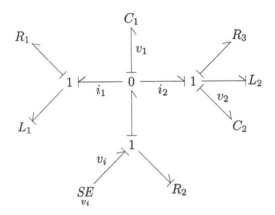

Figure S.1  The solution of Exercise 3.2.

**Exercise 3.3:** The correct answer is option **C)** because $\alpha$ and $\gamma$ represent 1 junctions. The complete bond graph solution is shown in Figure S.2.

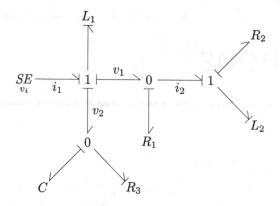

Figure S.2  The solution of Exercise 3.3.

**Exercise 3.4:** The correct answer is option **C)** since $\beta$ represents a 1 junction and $\delta$ a 0 junction. Figure S.3 shows the bond graph including all junctions.

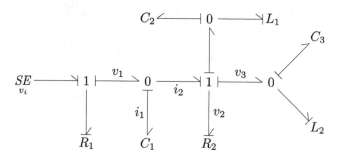

Figure S.3  The solution of Exercise 3.4.

**Exercise 3.5:** The correct answer is option **A)** as $\alpha = C_1$. Figure S.4 shows the bond graph including all symbols.

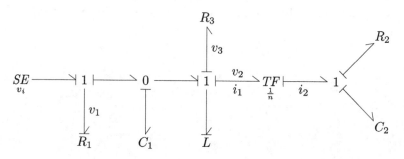

Figure S.4  The solution of Exercise 3.5.

## Exercise 3.6:

1. Figure S.5 shows the bond graph after assigning the causality, where no conflicts exist.

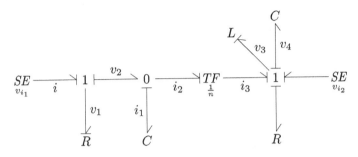

Figure S.5 The bond graph solution of Exercise 3.6.

2. Figure S.6 shows the complete electrical circuit of Exercise 3.6.

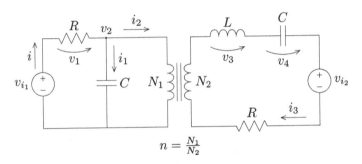

Figure S.6 The electrical circuit solution of Exercise 3.6.

**Exercise 3.7:** The electrical circuit derived from the bond graph of Exercise 3.7 is shown in Figure S.7.

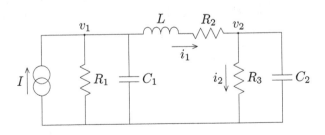

Figure S.7 The solution of Exercise 3.7.

**Exercise 3.8:** Figure S.8 shows the bond graph solution. We do not find any causal conflicts.

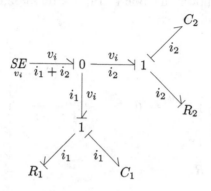

Figure S.8  The solution of Exercise 3.8.

**Exercise 3.9:** Figure S.9 shows the bond graph solution. We verify that it has no causality conflict.

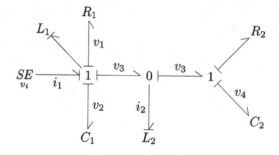

Figure S.9  The solution of Exercise 3.9.

**Exercise 3.10:** To completely specify the causality of the bond graph arbitrary causal strokes must be assigned to electrical resistances. In this case, we consider that $R_1$ and $R_2$ impose flow on their respective 1 junctions. The complete bond graph solution is shown in Figure S.10, where no causal conflict exist.

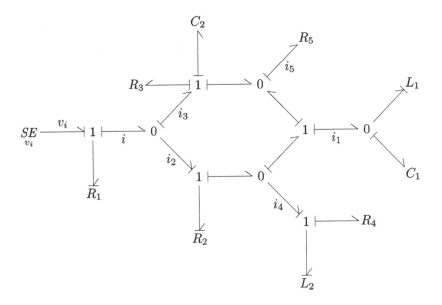

Figure S.10 The solution of Exercise 3.10.

**Exercise 3.11:** The complete bond graph is shown in Figure S.11. We do not find any conflict after applying the causality assignment rules.

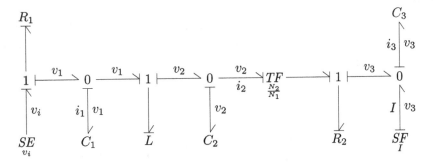

Figure S.11 The solution of Exercise 3.11.

**Exercise 3.12:** The complete bond graph is shown in Figure S.12, where no causal conflicts exist.

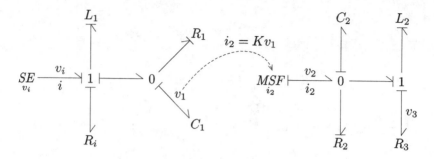

Figure S.12  The solution of Exercise 3.12.

## CHAPTER 4 SOLUTIONS

**Exercise 4.1:** The correct answer is option **A)** since $\alpha = B_1$, $\beta = B_2$, $\delta = M$, $\mu = K_2$ and $\varepsilon = K_1$.

**Exercise 4.2:** The correct answer is option **D)** given that $\alpha = M_1$, $\beta = M_2$ and $\gamma = K_1$.

**Exercise 4.3:** The correct answer is option **B)** because $\alpha = B_1$, $\beta = B_2$, $\delta = K_2$ and $\varepsilon = K_1$.

**Exercise 4.4:**

1. The correct answer is option **D)** since $\alpha = K_1$, $\beta = B_1$, $\gamma = M_1$, $\delta = K_2$, $\mu = M_2$, $\varepsilon = K_3$ and $\chi = M_3$. Figure S.13 shows the bond graph including all symbols.

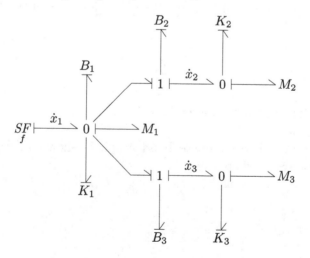

Figure S.13  The solution of Exercise 4.4.

2. The electrical circuit of Figure S.14 is the analogous of the mechanical system in Figure 4.23.

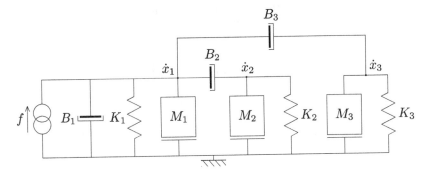

Figure S.14   The electrical circuit of Exercise 4.4.

## Exercise 4.5:

1. The correct answer is option **A**) as $\delta = M_2$ and $\varepsilon = M_3$.

2. The correct answer is option **A**) with $\beta = B_2$. The complete bond graph solution is shown in Figure S.15.

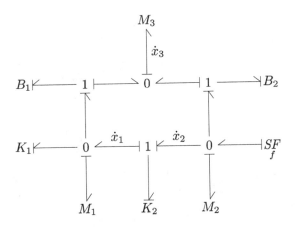

Figure S.15   The solution of Exercise 4.5.

## Exercise 4.6:

1. The correct junction types are $\alpha = 1$, $\beta = 0$ and $\gamma = 1$.

2. Figure S.16 shows the bond graph after assigning the causality, where no conflicts exist.

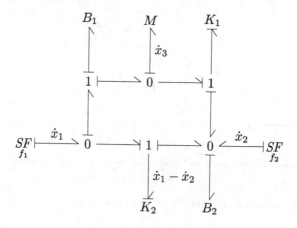

Figure S.16  The solution of Exercise 4.6.

3. The standard block diagram is depicted in Figure S.17.

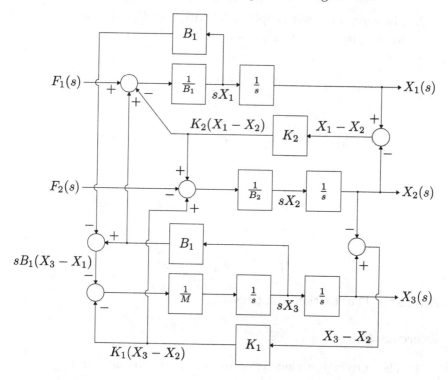

Figure S.17  The block diagram of Exercise 4.6 in the Laplace domain.

The classical mathematical model is:
$$\begin{cases} f_1 = B_1 \left( \dot{x}_1 - \dot{x}_3 \right) + K_2 \left( x_1 - x_2 \right) \\ 0 = M_3 \ddot{x}_3 + B_1 \left( \dot{x}_3 - \dot{x}_1 \right) + K_1 \left( x_3 - x_2 \right) \\ -f_2 = B_2 \dot{x}_2 + K_1 \left( x_2 - x_3 \right) + K_2 \left( x_2 - x_1 \right) \end{cases}$$

**Exercise 4.7:** The correct answer is option **A)** for the reason that $\gamma = B_1$. Figure S.18 shows the bond graph including the remaining symbols.

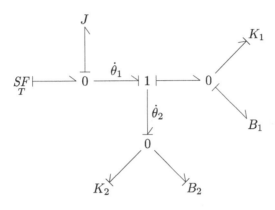

Figure S.18  The solution of Exercise 4.7.

**Exercise 4.8:**

1. The complete bond graph of Exercise 4.8 is shown in Figure S.19.

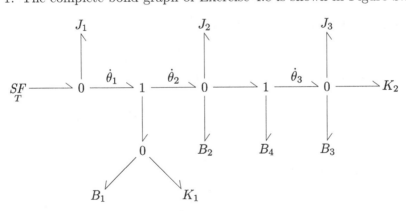

Figure S.19  The solution of Exercise 4.8.

2. The bond graph in Figure S.20 includes all causal relations, where no conflicts exist.

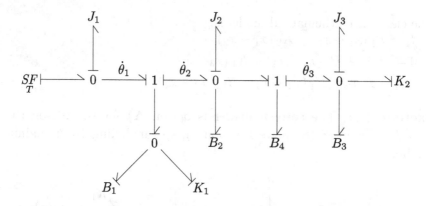

Figure S.20 The solution of Exercise 4.8 including all causal strokes.

**Exercise 4.9:** The correct answer is option **B)** given that $\alpha = B_1$. Figure S.21 shows the bond graph including all symbols.

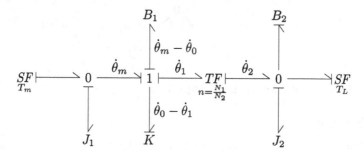

Figure S.21 The solution of Exercise 4.9.

**Exercise 4.10:** The bond graph solution of Exercise 4.10 is shown in Figure S.22. It has no causality conflicts.

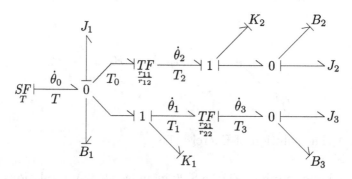

Figure S.22 The solution of Exercise 4.10.

**Exercise 4.11:** A possible rotational mechanical system derived from the bond graph of Exercise 4.11 is shown in Figure S.23, where the causal conflict was address by adding elasticity ($K'$) to the axis that connects the belt drive to the inertia $J_2$. Note that the $TF$ element can be either implemented by means of a belt drive or a gear, given that both have the same primary and secondary power variables relations, $\frac{T_1}{T_2} = \frac{\theta_2}{\theta_1} = \frac{r_1}{r_2}$. The updated bond graph of the exercise is shown in Figure S.24.

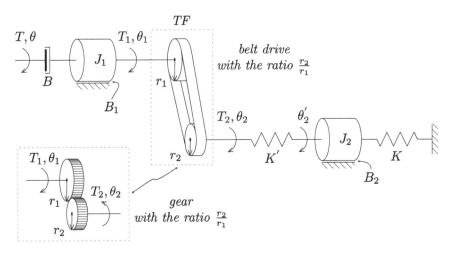

Figure S.23 The proposed mechanical system for Exercise 4.11, with the causality conflict solved.

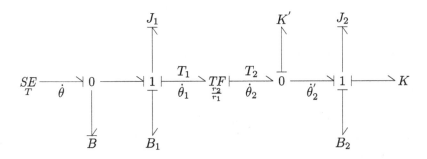

Figure S.24 The Option 2 bond graph of the system in Figure S.23.

**Exercise 4.12:** Figure S.25 shows the bond graph solution. We do not find any causality conflict.

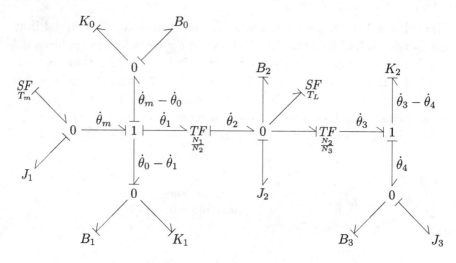

Figure S.25  The solution of Exercise 4.12.

**Exercise 4.13:** The correct answer is option **B)** since $\alpha = M$. The complete bond graph is shown in Figure S.26.

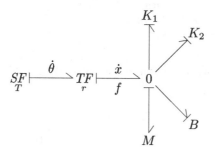

Figure S.26  The bond graph of Exercise 4.13.

**Exercise 4.14:**

1. The correct answer is option **A)** since the symbol $\alpha$ denotes two 0 junctions. Figure S.27 shows the bond graph solution including all symbols and causality.

2. See Figure S.27.

3. The correct answer is option **A)**. The bond graph has a causal conflict in the 0 junction that connects the elements $TF$, $SF$ and $M$.

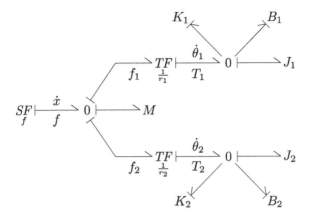

Figure S.27   The solution of Exercise 4.14.

**Exercise 4.15:**

1. (a) The correct answer is option **A)** given that $\beta = 1$ and $\delta = B_1$.
   (b) The correct answer is option **C)** since $\alpha = K_2$ and $\gamma = 0$. Figure S.28 shows the bond graph solution including all symbols.

2. Figure S.28 shows the bond graph with the causality assigned.

3. The bond graph has no causality conflict, thus the correct answer is option **B)**.

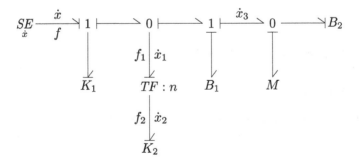

Figure S.28   The solution of Exercise 4.15.

**Exercise 4.16:**   Figure S.29 shows the bond graph solution. We do not find any causal conflict.

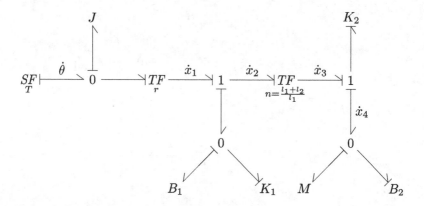

Figure S.29  The solution of Exercise 4.16.

**Exercise 4.17:** The correct answer is option **C)** as $\alpha = M_L$. Figure S.30 shows the bond graph solution including all symbols.

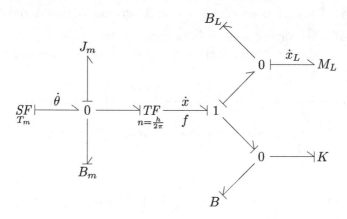

Figure S.30  The solution of Exercise 4.17.

**Exercise 4.18:**

1. The correct answer is option **C)** given that $\zeta = K$ and $\varepsilon = B_2$. Figure S.31 shows the bond graph solution including all symbols and causality.

2. It can be observed from Figure S.31 that the bond graph has no causality conflict, therefore, the correct answer is option **B)**.

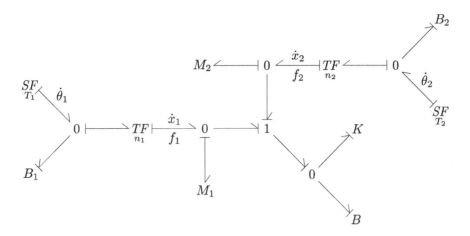

Figure S.31 The solution of Exercise 4.18.

**Exercise 4.19:**

1. The correct answer is option **A)** for the reason that $\alpha = \dot{x}_1$, $\beta = f$, $\gamma = 0$, $\delta = 1$ and $\varepsilon = 0$.

2. Figure S.32 shows the bond graph solution including all symbols and causality. It has no causal conflicts.

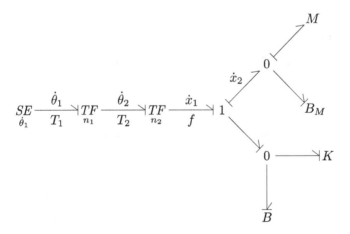

Figure S.32 The solution of Exercise 4.19.

**Exercise 4.20:** Figure S.33 shows the bond graph solution. We do not find any causality conflict.

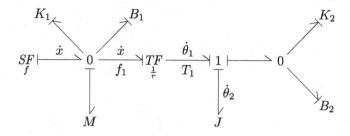

Figure S.33  The solution of Exercise 4.20.

**Exercise 4.21:** Figure S.34 shows the bond graph solution. We do not find any causality conflicts.

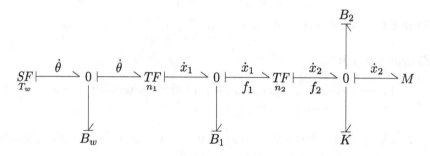

Figure S.34  The solution of Exercise 4.21.

## CHAPTER 5 SOLUTIONS

**Exercise 5.1:** The correct answer is option **B)** given that $\alpha = C_1$, $\beta = SF$, $\gamma = C_2$, $\delta = SF$ and $\varepsilon = R$.

**Exercise 5.2:** The correct answer is option **A)** since $\alpha = C_1$. Figure S.35 shows the bond graph including all symbols.

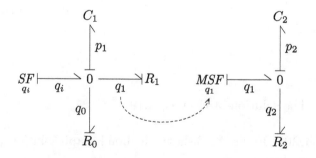

Figure S.35  The solution of Exercise 5.2.

**Exercise 5.3:** The correct answer is option **D**), where $\alpha = C_1$, $\beta = C_2$ and $\gamma = C_3$. Figure S.36 shows the bond graph including all symbols.

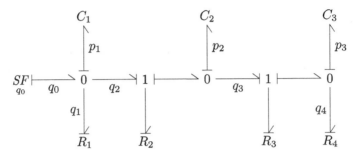

Figure S.36  The solution of Exercise 5.3.

**Exercise 5.4:** The correct answer is option **C**) since $\alpha = 0$ and $\gamma = 0$. Figure S.37 shows the bond graph including all symbols and causality.

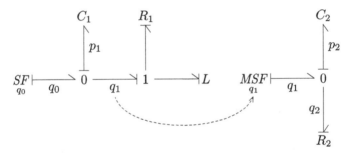

Figure S.37  The solution of Exercise 5.4.

**Exercise 5.5:** Figure S.38 shows the bond graph solution. We do not find any causality conflict.

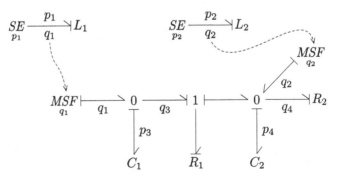

Figure S.38  The solution of Exercise 5.5.

**Exercise 5.6:** The correct answer is option **B)** given that $\alpha = C_1$, $\beta = R_1$, $\gamma = C_2$ and $\delta = R_2$.

**Exercise 5.7:**

1. The correct answer is option **C)** as $\gamma = R_2$. Figure S.39 shows the bond graph including all symbols.

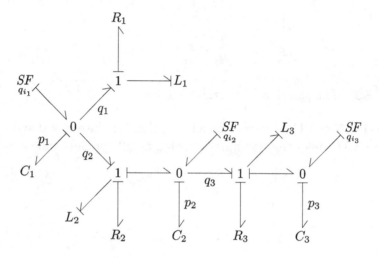

Figure S.39  The solution of Exercise 5.7.

2. Figure S.40 shows the analogous electrical circuit of the hydraulic system of Figure 5.24.

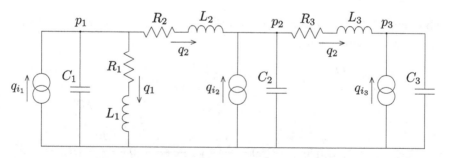

Figure S.40  The electrical circuit of Exercise 5.7.

**Exercise 5.8:** The correct answer is option **A)** since $\alpha = R_1$. Figure S.41 shows the bond graph including all symbols.

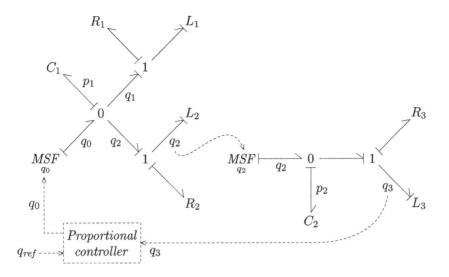

Figure S.41  The solution of Exercise 5.8.

**Exercise 5.9:** Figure S.42 shows the bond graph solution. We verify that there is a causal conflict in the 1 junction that connects the elements *MSF*, $R_3$ and $L$.

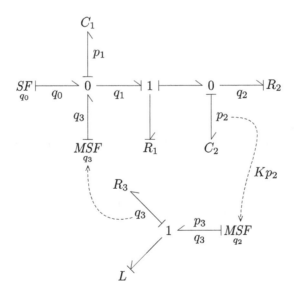

Figure S.42  The solution of Exercise 5.9.

**Exercise 5.10:** Figure S.43 shows the bond graph solution and it has no causal conflict.

## 216 ■ Solutions

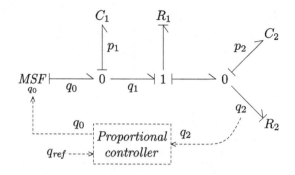

Figure S.43 The solution of Exercise 5.10.

**Exercise 5.11:** One possible hydraulic system is shown in Figure S.44.

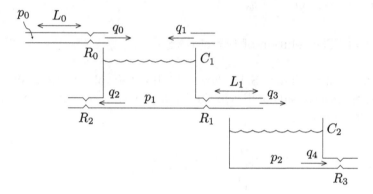

Figure S.44 The solution of Exercise 5.11.

**Exercise 5.12:** Figure S.45 shows the bond graph solution. We verify that there is a causal conflict in the bond that connects the elements $MSF$ and $L$.

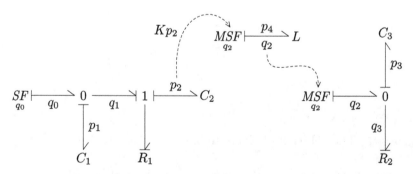

Figure S.45 The solution of Exercise 5.12.

**Exercise 5.13:** Figure S.46 shows the bond graph solution. We do not find any causality conflict.

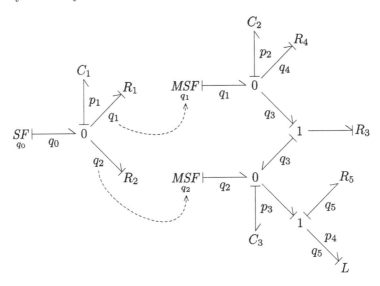

Figure S.46 The solution of Exercise 5.13.

**Exercise 5.14:**

1. The dynamic equations of Exercise 5.14, considering zero initial conditions:

$$\begin{cases} q_{i1} = q_1 + q_2 + q_{C_1}, \; q_{i2} + q_2 = q_3 + q_{C_2}, \; q_1 + q_3 = q_{C_3} \\ q_1 = \frac{1}{L_1} \int_0^t p_{L_1}(\tau) \, d\tau \\ q_2 = \frac{1}{L_2} \int_0^t p_{L_2}(\tau) \, d\tau \\ q_3 = \frac{1}{L_3} \int_0^t p_{L_3}(\tau) \, d\tau \\ p_1 = \frac{1}{C_1} \int_0^t q_{C_1}(\tau) \, d\tau \\ p_2 = \frac{1}{C_2} \int_0^t q_{C_2}(\tau) \, d\tau \\ p_3 = \frac{1}{C_3} \int_0^t q_{C_3}(\tau) \, d\tau \\ p_1 = R_1 q_1 + p_{L_1}, \; p_2 = R_3 q_3 + p_{L_3} \\ p_1 - p_2 = R_2 q_2 + p_{L_2} \end{cases}$$

2. Figure S.47 shows the bond graph solution. It has no causal conflicts.

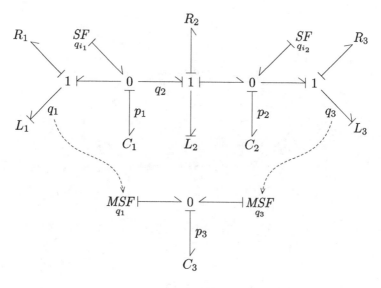

Figure S.47  The solution of Exercise 5.14.

**Exercise 5.15:** The bond graph solution is shown in Figure S.48 and it is causal conflict free.

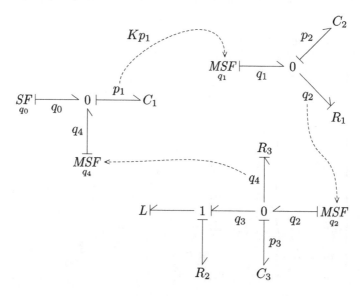

Figure S.48  The solution of Exercise 5.15.

**Exercise 5.16:** Figure S.49 shows the bond graph solution. We do not find any causality conflict.

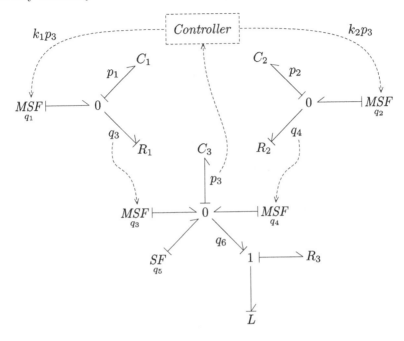

Figure S.49  The solution of Exercise 5.16.

## CHAPTER 6 SOLUTIONS

**Exercise 6.1:** The correct answer is option **A)** since $\alpha = 0$, $\beta = 1$, $\gamma = C_2$ and $\delta = R$. The complete bond graph solution is shown in Figure S.50.

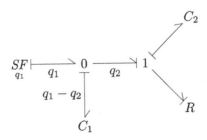

Figure S.50  The solution of Exercise 6.1.

**Exercise 6.2:** The correct answer is option **C)** given that $\alpha = 1$, $\beta = 0$ and $\varepsilon = R_2$. The complete bond graph solution is shown in Figure S.51.

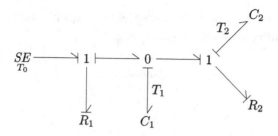

Figure S.51 The solution of Exercise 6.2.

**Exercise 6.3:** The correct answer is option **A)** since $\alpha = C_1$ and $\beta = C_2$. Figure S.52 shows the bond graph including all symbols.

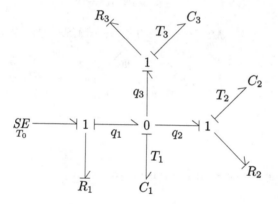

Figure S.52 The solution of Exercise 6.3.

**Exercise 6.4:** The correct answer is option **D)** since $\alpha = R_{pf}$, $\gamma = C_g$ and $\varepsilon = R_{am}$. Figure S.53 shows the bond graph including all symbols.

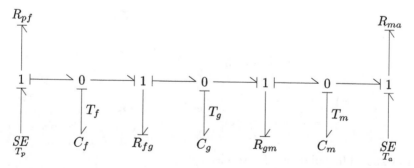

Figure S.53 The solution of Exercise 6.4.

## Exercise 6.5:

1. Figure S.54 shows the bond graph after assigning the causality, where no conflicts exist.

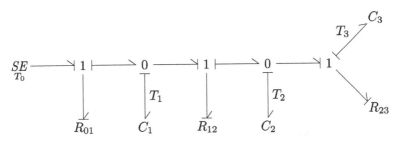

Figure S.54 The bond graph solution of Exercise 6.5.

2. We consider the existence of a perfectly insulated reservoir with 'material 0', that is at a constant temperature $T_0$, and three other bodies called 'material 1', 'material 2' and 'material 3'. These are at temperatures $T_1$, $T_2$ and $T_3$, and have the thermal capacitances $C_1$, $C_2$ and $C_3$. Moreover, we consider that $R_{01}$, $R_{12}$ and $R_{23}$ represent the thermal resistances to the heat flows between the different materials. A possible thermal system derived from the bond graph of Exercise 6.5 is shown in Figure S.55.

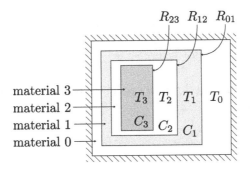

Figure S.55 The thermal system of Exercise 6.5.

**Exercise 6.6:** Figure S.56 presents the bond graph of Exercise 6.6. We do not find any causality conflicts.

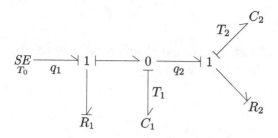

Figure S.56  The solution of Exercise 6.6.

**Exercise 6.7:** The complete bond graph is shown in Figure S.57. We do not find any conflict with the causal relations.

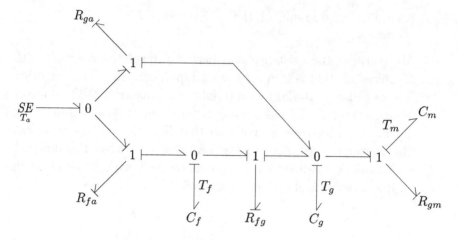

Figure S.57  The solution of Exercise 6.7.

**Exercise 6.8:** The bond graph of the exercise is shown in Figure S.58. It has no causality conflicts.

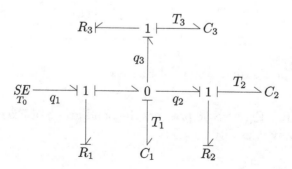

Figure S.58  The solution of Exercise 6.8.

Solutions ■ 223

**Exercise 6.9:** Figure S.59 shows the bond graph solution. We do not find any causality conflicts.

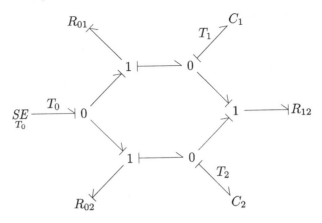

Figure S.59  The solution of Exercise 6.9.

**Exercise 6.10:**

1. To complete the causality assignment some arbitrary causal strokes must be given to thermal resistances. Figure S.60 shows the bond graph after assigning the causality, where no conflicts exist.

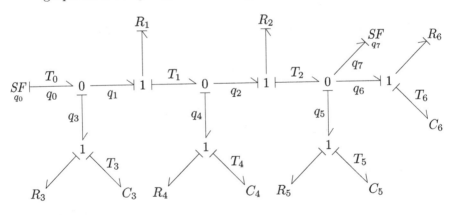

Figure S.60  The solution of Exercise 6.10.

2. The mathematical model of Exercise 6.10 is (considering zero initial conditions):

$$\begin{cases} q_0 = q_1 + q_3, \; q_1 = q_2 + q_4, \; q_2 = q_5 + q_6 + q_7 \\ \frac{T_0-T_1}{R_1} = q_1, \; \frac{T_1-T_2}{R_2} = q_2, \; \frac{T_0-T_3}{R_3} = q_3 \\ \frac{T_1-T_4}{R_4} = q_4, \; \frac{T_2-T_5}{R_5} = q_5, \; \frac{T_2-T_6}{R_6} = q_6 \\ T_3 = \frac{1}{C_3} \int_0^t q_3(\tau)\,d\tau, \; T_4 = \frac{1}{C_4} \int_0^t q_4(\tau)\,d\tau \\ T_5 = \frac{1}{C_5} \int_0^t q_5(\tau)\,d\tau, \; T_6 = \frac{1}{C_6} \int_0^t q_6(\tau)\,d\tau \end{cases}$$

The standard block diagram is depicted in Figure S.61.

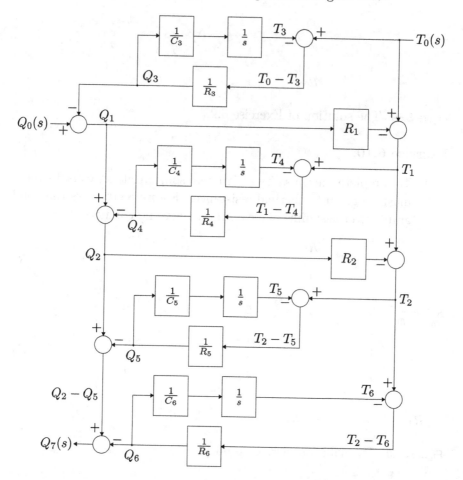

Figure S.61 The block diagram of Exercise 6.10 in the Laplace domain.

**Exercise 6.11:** Figure S.62 shows the bond graph solution. We do not find any causality conflicts.

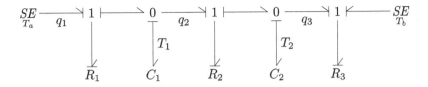

Figure S.62   The solution of Exercise 6.11.

**Exercise 6.12:**

1. Figure S.63 shows the bond graph after assigning the causality. We do not identify any causal conflict.

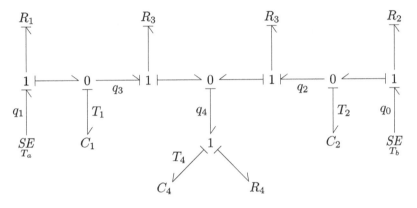

Figure S.63   The solution of Exercise 6.12.

2. Figure S.64 shows the electrical analogous circuit of the thermal system of Exercise 6.12.

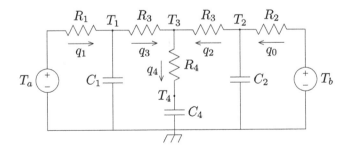

Figure S.64   The electrical circuit of Exercise 6.12 using the {temperature, heat flow rate}→{voltage, current} analogy.

**Exercise 6.13:** Figure S.65 shows the bond graph solution. We do not find any causality conflicts.

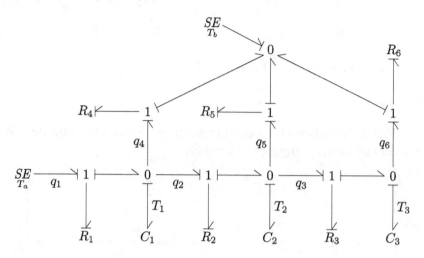

Figure S.65  The solution of Exercise 6.13.

**Exercise 6.14:** Figure S.66 shows the bond graph solution. We do not find any causality conflicts.

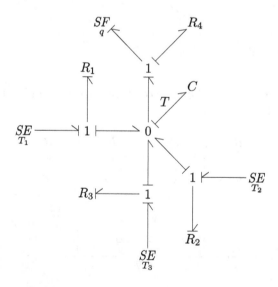

Figure S.66  The solution of Exercise 6.14.

# CHAPTER 7 SOLUTIONS

**Exercise 7.1:**

1. The correct answer is option **C)** given that $\alpha = B$, $\beta = M$ and $\delta = K$.

2. The correct answer is option **B)** since we do not find any causal conflict in the bond graph.

**Exercise 7.2:** The correct answer is option **B)** since $\alpha = 0$. Figure S.67 shows the bond graph including all symbols.

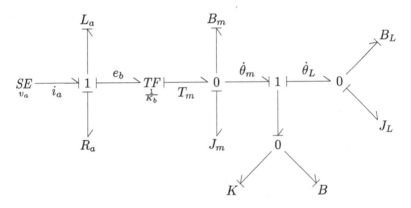

Figure S.67  The solution of Exercise 7.2.

**Exercise 7.3:** The bond graph of the exercise is shown in Figure S.68. It has no causality conflict.

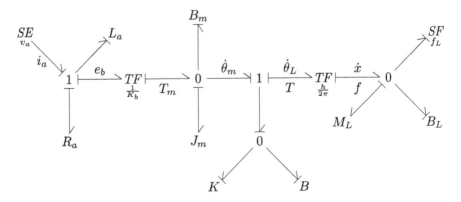

Figure S.68  The solution of Exercise 7.3.

## Exercise 7.4:

1. The correct answer is option **A)** given that $\alpha = \omega_1$ and $\beta = K_2$. Moreover, the bond graph of Figure S.69 shows all symbols.

2. Figure S.69 depicts the bond graph after assigning the causality, where we do not find any conflict.

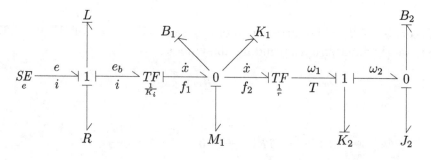

Figure S.69 The solution of Exercise 7.4.

## Exercise 7.5:

1. Figure S.70 shows the bond graph solution for Exercise 7.5, where we do not find any causality conflicts.

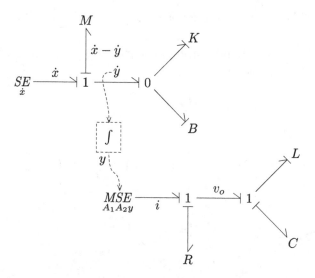

Figure S.70 The solution of Exercise 7.5.

2. The mathematical model of Exercise 7.5 is (zero initial conditions):
$$\begin{cases} M(\ddot{x} - \ddot{y}) + B\dot{y} + Ky = 0 \\ v_i = A_1 y \\ v_2 = A_2 v_i = A_1 A_2 y \\ v_2 = Ri + v_L + \frac{1}{C} \int_0^t i(\tau) \, d\tau \\ i = \frac{1}{L} \int_0^t v_L(\tau) \, d\tau \\ v_o = v_L + \frac{1}{C} \int_0^t i(\tau) \, d\tau \end{cases}$$

Figure S.71 depicts the standard block diagram.

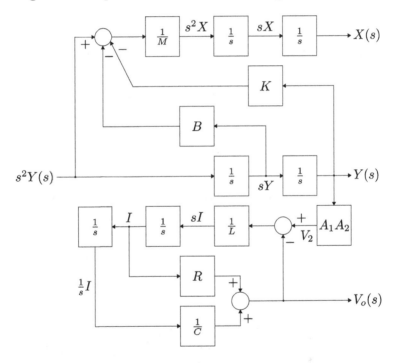

Figure S.71 The block diagram of Exercise 7.5 in the Laplace domain.

**Exercise 7.6:** From the equations comes that $\frac{q_0}{\epsilon A}$ is the coupling factor between the electrical and mechanical parts of the system. The complete bond graph is shown in Figure S.72, where no causal conflict exist.

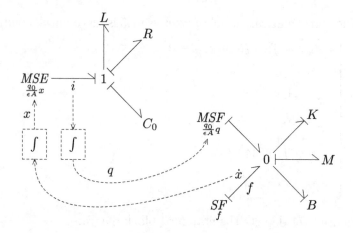

Figure S.72 The solution of Exercise 7.6.

The standard block diagram is represented in Figure S.73.

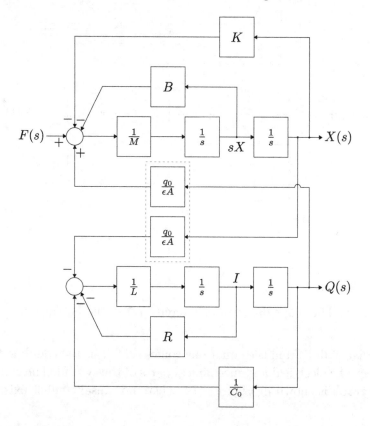

Figure S.73 The block diagram of Exercise 7.6 in the Laplace domain.

## Exercise 7.7:

1. The correct answer is option **C)**, since $\alpha = 1$, $\beta = 0$, $\gamma = 0$, $\delta = GY$ and $\varepsilon = 1$.

2. Figure S.74 shows the bond graph solution for Exercise 7.7 after assigning the causality. We do not find any conflict.

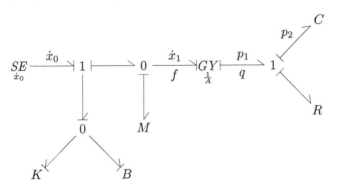

Figure S.74  The solution of Exercise 7.7.

3. Adopting the {pressure, flow rate}→{voltage, current} correspondence for the hydraulic subsystem and the {velocity, force}→{voltage, current} analogy for the mechanical part, we can derive the analogous electrical circuit shown in Figure S.75.

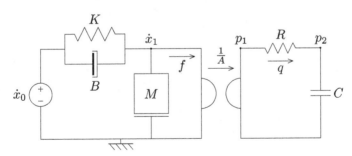

Figure S.75  The electrical circuit of Exercise 7.7.

**Exercise 7.8:** Figure S.76 shows the complete bond graph of Exercise 7.8, where we do not find any causality conflict.

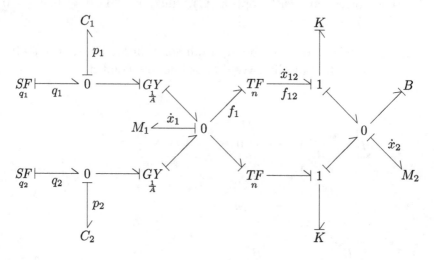

Figure S.76 The solution of Exercise 7.8.

**Exercise 7.9:** The complete bond graph of Exercise 7.9 is shown in Figure S.77. We do not find any conflict with the causality.

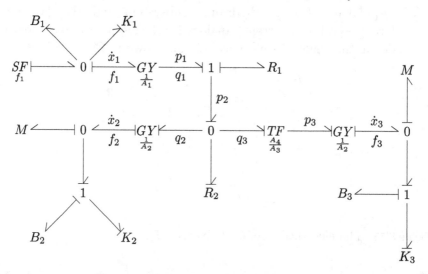

Figure S.77 The solution of Exercise 7.9.

Solutions ■ 233

**Exercise 7.10:** A possible system derived from the bond graph of Exercise 7.10 is shown in Figure S.78.

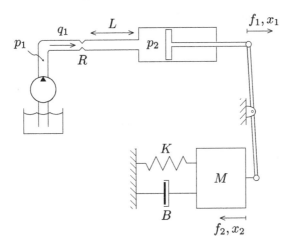

Figure S.78  The hydraulic-mechanical system for Exercise 7.10.

One possible solution for the causal conflict is to add elasticity to the connection between the lever and the mechanical subsystem, like proposed in the system of Figure S.79. The resulting changes in the bond graph of the exercise are shown in Figure S.80.

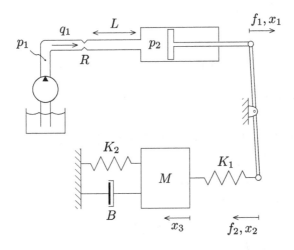

Figure S.79  One of the possible solutions for the causality conflict of Exercise 7.10.

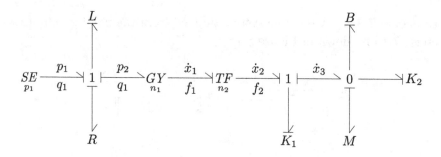

Figure S.80  The bond graph of the multi-domain system of Figure S.79.

## Exercise 7.11:

1. Figure S.81 shows the bond graph solution for Exercise 7.11, where we do not find any causal conflicts.

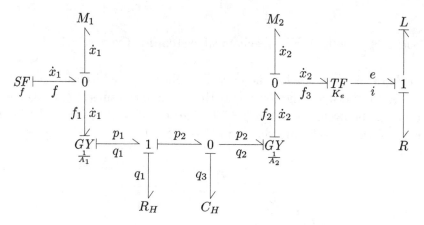

Figure S.81  The solution of Exercise 7.11.

2. The mathematical model of Exercise 7.11 is:
$$\begin{cases} f = f_1 + M_1\ddot{x}_1, \ f_2 = f_3 + M_2\ddot{x}_2 \\ f_1 = A_1 p_1, \ f_2 = A_2 p_2 \\ \dot{x}_1 A_1 = q_1, \ \dot{x}_2 A_2 = q_2 \\ \frac{p_1 - p_2}{R_H} = q_1, \ q_1 = q_2 + q_3, q_3 = C_H \dot{p}_2 \\ f_3 = K_e i \\ e = K_e \dot{x}_2, \ e = Ri + L\frac{di}{dt} \end{cases}$$

The system's standard block diagram is shown in Figure S.82.

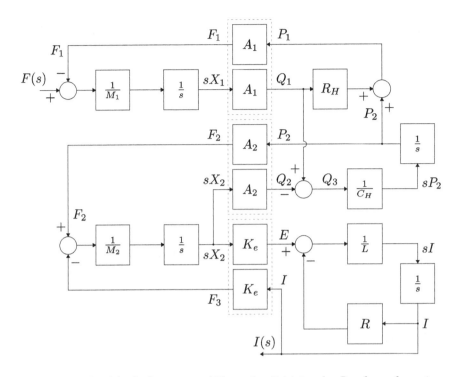

Figure S.82  The block diagram of Exercise 7.11 in the Laplace domain.

**Exercise 7.12:** The bond graph in Figure S.83 shows the solution, where no causal conflict exist.

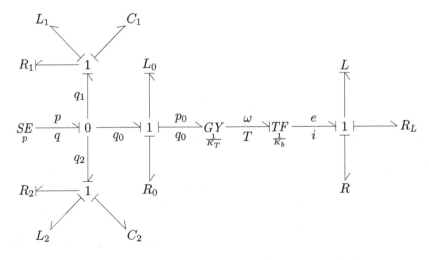

Figure S.83  The solution of Exercise 7.12.

**Exercise 7.13:** Figure S.84 shows the bond graph solution for Exercise 7.13, where we do not find any causality conflict.

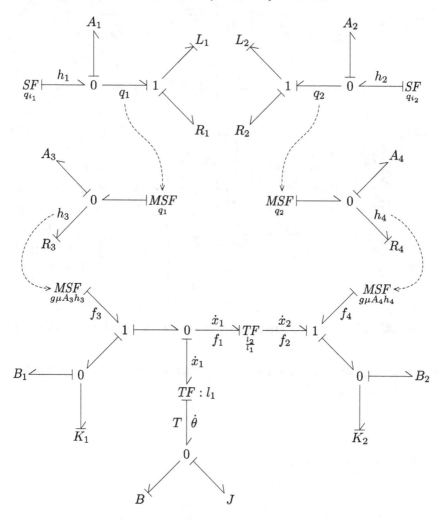

Figure S.84  The solution of Exercise 7.13.

**Exercise 7.14:** The bond graph in Figure S.85 shows the solution that covers four energy domains. We consider it starts with the temperature source provided by the reactor that enables the production of hydraulic gas. The steam pressure propels the turbine ($TF$ with the transformation ratio $n_1$) which outputs rotational mechanical power that drives the electrical generator. This device performs the conversion from mechanical to electrical power with the ratio $n_2$. Finally, the electrical transformer using the ratio $n_3$ supplies the electrical grid represented by $R_L$. The sub-process that maintains a constant steam generator water height, and the one that keeps a constant condenser water temperature were not considered for the bond graph modeling. The solution has no causal conflict.

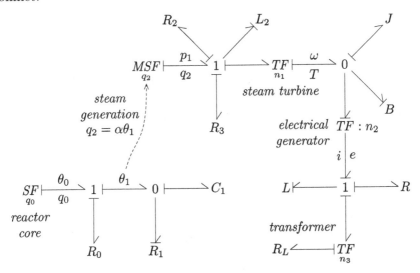

Figure S.85  The solution of Exercise 7.14.

**Exercise 7.15:** The solution of Exercise 7.15 has two bond graphs. One models the pressure controlling process (Figure S.86), while the bond graph of Figure S.87 models the temperature control process. Both have no causality conflicts.

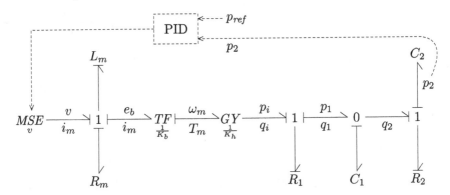

Figure S.86  The solution of Exercise 7.15, not including the thermal subsystem.

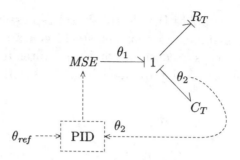

Figure S.87  The temperature control subsystem solution of Exercise 7.15.

**Exercise 7.16:** Figure S.88 shows the bond graph solution. It can be observed that the graph has no causal conflict.

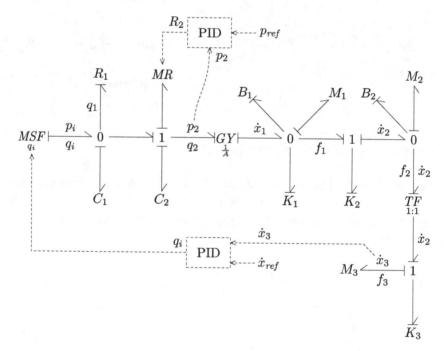

Figure S.88  The solution of Exercise 7.16.

**Exercise 7.17:**

1. The bond graph of the human body as a mechanical system is shown in Figure S.89. No causal conflicts exist.

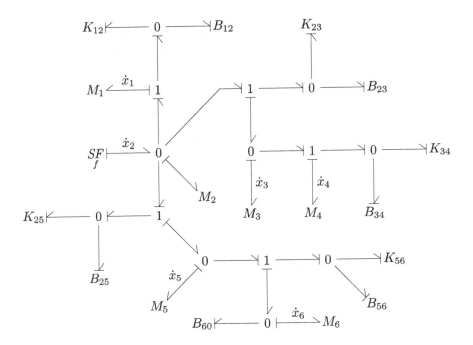

Figure S.89 The solution of question number 1 of Exercise 7.17.

2. Figure S.90 shows the electrical analogous circuit of the mechanical system in Figure 7.32.

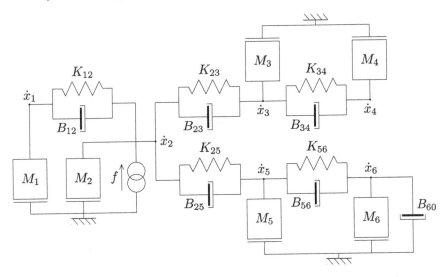

Figure S.90 The electrical circuit of question number 2 of Exercise 7.17.

**240** ◼ Solutions

3. The human body as a thermal system results on the bond graph in Figure S.91, where we do not find any causal conflict.

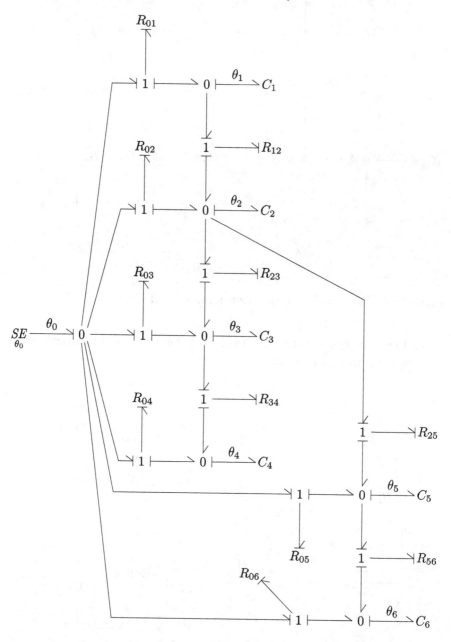

Figure S.91  The solution of question number 3 of Exercise 7.17.

4. Figure S.92 shows the electrical analogous circuit of the thermal system in Figure 7.33.

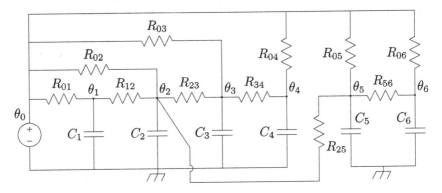

Figure S.92 The electrical circuit of question number 4 of Exercise 7.17.

## Exercise 7.18:

1. The bond graph of the plant leaf as a mechanical system is shown in Figure S.93, where no causal conflict exist.

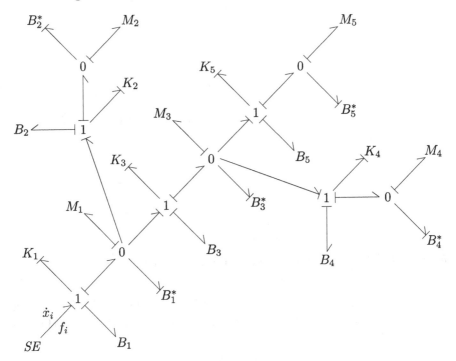

Figure S.93 The solution of question number 1 of Exercise 7.18.

2. Adopting the provided hydraulic analogy and applying it to the five sections of the leaf we obtain the bond graph shown in Figure S.94. Additionally, after assigning the causal strokes we verify that the bond graph is causal conflict free.

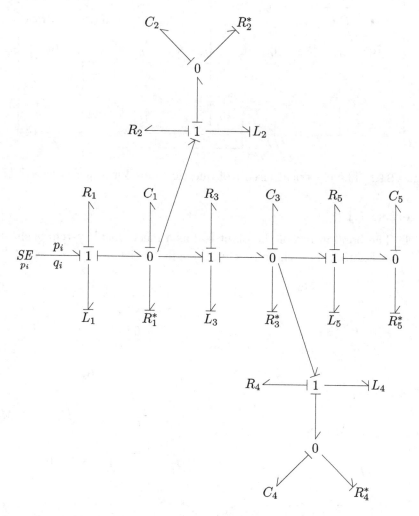

Figure S.94  The solution of question number 2 of Exercise 7.18.

3. The electrical circuit in Figure S.95 shows the result of adopting the {pressure, flow rate}→{voltage, current} analogy to the hydraulic system of Figure 7.35.

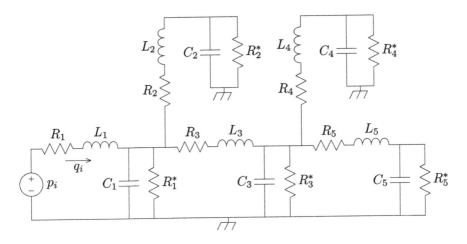

Figure S.95  The electrical circuit of question number 3 of Exercise 7.18.

# Appendix A

TABLE A.1 Fundamental variables, units and symbols common to all systems.

| Time | $t$ | second | s |
|---|---|---|---|
| Energy | $W$ | joule | J |
| Power | $P$ | watt | W |

TABLE A.2 Fundamental variables and basic elements, units and symbols for electrical systems.

| Magnetic flux | $\lambda$ | weber | Wb |
|---|---|---|---|
| Charge | $q$ | coulomb | C |
| Current | $i$ | ampere | A |
| Voltage | $v$ | volt | V |
| Resistance | $R$ | ohm | $\Omega$ |
| Inductance | $L$ | henry | H |
| Capacitance | $C$ | farad | F |

TABLE A.3 Fundamental variables and basic elements, units and symbols for translational mechanical systems.

| Momentum | $p$ | kilogram meter per second | kg · m · s$^{-1}$ |
|---|---|---|---|
| Displacement | $x$ | meter | m |
| Force | $f$ | newton | N |
| Velocity | $v$ | meter per second | m · s$^{-1}$ |
| Damping coefficient | $B$ | newton second per meter | N · s · m$^{-1}$ |
| Stiffness coefficient | $K$ | newton per meter | N · m$^{-1}$ |
| Mass | $M$ | kilogram | kg |

**TABLE A.4** Fundamental variables and basic elements, units and symbols for rotational mechanical systems.

| | | | |
|---|---|---|---|
| Momentum | $\mathcal{M}$ | kilogram squared meter per second | $kg \cdot m^2 \cdot s^{-1}$ |
| Displacement | $\theta$ | radian | rad |
| Torque | $T$ | newton meter | $N \cdot m$ |
| Velocity | $\omega$ | radians per second | $rad \cdot s^{-1}$ |
| Damping coefficient | $B$ | newton meter second per radian | $N \cdot m \cdot s \cdot rad^{-1}$ |
| Stiffness coefficient | $K$ | newton meter per radian | $N \cdot m \cdot rad^{-1}$ |
| Inertia | $J$ | kilogram squared meter | $kg \cdot m^2$ |

**TABLE A.5** Fundamental variables and basic elements, units and symbols for hydraulic systems.

| | | | |
|---|---|---|---|
| Volume | $V$ | cubic meter | $m^3$ |
| Fluid momentum | $\Gamma$ | pascal second | $Pa \cdot s$ |
| Volumetric flow rate | $q$ | cubic meter per second | $m^3 \cdot s^{-1}$ |
| Pressure | $p$ | pascal | Pa |
| Resistance | $R$ | pascal second per cubic meter | $Pa \cdot s \cdot m^{-3}$ |
| Inertance | $L$ | pascal squared second per cubic meter | $Pa \cdot s^2 \cdot m^{-3}$ |
| Capacitance | $C$ | cubic meter pascal | $m^3 \cdot Pa$ |

**TABLE A.6** Fundamental variables and basic elements, units and symbols for thermal systems.

| | | | |
|---|---|---|---|
| Total heat | $H$ | joule | J |
| Heat flow rate | $q$ | watt | W |
| Temperature | $T$ | kelvin | K |
| Resistance | $R$ | kelvin per watt | $K \cdot W^{-1}$ |
| Capacitance | $C$ | joule per kilogram kelvin | $J \cdot Kg^{-1} \cdot K^{-1}$ |

**TABLE A.7** Bond graph elements causalities.

| Elements type | | |
|---|---|---|
| Sources (generalized) | effort<br>$SE \;\vdash\!\!\!\longrightarrow$<br>$\;\;\;e$ | flow<br>$SF \vdash\!\!\longrightarrow$<br>$\;\;\;f$ |
| Stores (generalized) | effort<br>$\longrightarrow\!\!\vdash L$ | flow<br>$\vdash\!\!\longrightarrow C$ |
| Dissipator (generalized) | $\longrightarrow\!\!\vdash R$ | $\vdash\!\!\longrightarrow R$ |
| Junctions | 0 junction<br><br>$\begin{array}{c} f_2 \mid e \\ \downarrow \\ \dfrac{e}{f_1}\!\longrightarrow\!\! 0 \!\!\longleftarrow\!\! \dfrac{e}{f_3} \\ \uparrow \\ f_n \mid e \quad \cdots \end{array}$ | 1 junction<br><br>$\begin{array}{c} f \mid e_2 \\ \downarrow \\ \dfrac{e_1}{f}\!\vdash\!\! 1 \!\!\dashv\! \dfrac{e_3}{f} \\ \uparrow \\ f \mid e_n \quad \cdots \end{array}$ |
| Coupling | transformer<br>$\vdash\!\!\longrightarrow TF\vdash\!\!\longrightarrow$<br>$\longrightarrow\!\!\dashv TF\longrightarrow\!\!\dashv$ | gyrator<br>$\longrightarrow\!\!\dashv GY\vdash\!\!\longrightarrow$<br>$\vdash\!\!\longrightarrow GY\longrightarrow\!\!\dashv$ |

**TABLE A.8** Accumulation elements causality by physical domain.

| Domain | Effort store | Flow store | |
|---|---|---|---|
| Electrical | $\longrightarrow\!\!\dashv L$ | $\vdash\!\!\longrightarrow C$ | |
| Mechanical (translational) | $\longrightarrow\!\!\dashv K$ | $\vdash\!\!\longrightarrow M$ | Option 1 |
| Mechanical (rotational) | $\longrightarrow\!\!\dashv K$ | $\vdash\!\!\longrightarrow J$ | Option 1 |
| Mechanical (translational) | $\longrightarrow\!\!\dashv M$ | $\vdash\!\!\longrightarrow K$ | Option 2 |
| Mechanical (rotational) | $\longrightarrow\!\!\dashv J$ | $\vdash\!\!\longrightarrow K$ | Option 2 |
| Hydraulic | $\longrightarrow\!\!\dashv L$ | $\vdash\!\!\longrightarrow C$ | |
| Thermal | no equivalent | $\vdash\!\!\longrightarrow C$ | |

# Appendix B

In order to model electrical systems with the 20-sim software, consider the equivalences between generalized bond graph elements and 20-sim library elements listed in Table B.1. The 20-sim software application adopts the 'Option 2' analogy for mechanical systems {force/torque, linear/angular velocity}→{effort, flow}. Refer to Table B.2 or Table B.3 to find the correct elements to use when adopting the 'Option 1' or 'Option 2' analogy, respectively. Although these tables only refer to rotational mechanical systems, the same applies to translational mechanical systems. For hydraulic and thermal systems modeling consider the Table B.4 and Table B.5, respectively. Table B.6 lists the 20-sim transducer library element to use according to the adopted analogy option.

TABLE B.1 Electrical systems elements and the equivalent elements in 20-sim.

| Element | Electrical domain | 20-sim library element |
|---|---|---|
| Effort source | Voltage source $(v)$ | Se |
| Flow source | Current source $(i)$ | Sf |
| Effort store | Inductor $(L)$ | I |
| Flow store | Capacitor $(C)$ | C |
| Dissipator | Resistor $(R)$ | R |

TABLE B.2 Option 1 (rotational) mechanical elements and the equivalent elements in 20-sim.

| Element | Mechanical domain (Option 1) | 20-sim library element |
|---|---|---|
| Effort source | Velocity source $(\omega)$ | Se |
| Flow source | Torque source $(T)$ | Sf |
| Effort store | Torsional spring $(K)$ | I |
| Flow store | Inertia $(J)$ | C |
| Dissipator | Rotational friction/damper $(B)$ | R with the equation $p.e = (1/r) * p.f$ |

**TABLE B.3** Option 2 (rotational) mechanical elements and the equivalent elements in 20-sim.

| Element | Mechanical domain (Option 2) | 20-sim library element |
|---|---|---|
| Effort source | Torque source ($T$) | Se |
| Flow source | Velocity source ($\omega$) | Sf |
| Effort store | Inertia ($J$) | I |
| Flow store | Torsional spring ($K$) | C |
| Dissipator | Rotational friction/damper ($B$) | R |

**TABLE B.4** 20-sim library elements for hydraulic systems.

| Element | Hydraulic domain | 20-sim library element |
|---|---|---|
| Effort source | Pressure source ($p$) | Se |
| Flow source | Fluid flow source ($q$) | Sf |
| Effort store | Inertance ($L$) | I |
| Flow store | Reservoir Capacitance ($C$) | C |
| Dissipator | Hydraulic resistance ($R$) | R |

**TABLE B.5** 20-sim library elements for thermal systems.

| Element | Thermal domain | 20-sim library element |
|---|---|---|
| Effort source | Temperature source ($T$) | Se |
| Flow source | Heat flow source ($q$) | Sf |
| Flow store | Material thermal capacity ($C$) | C |
| Dissipator | Fourier law | R |

**TABLE B.6** Most common transducer elements and the 20-sim element to use depending on the adopted analogy Option.

| Transducer | 20-sim element (Option 1) | 20-sim element (Option 2) |
|---|---|---|
| Electrical transformer | TF | TF |
| DC motor | TF | GY |
| Mechanical lever, gear and belt drive | TF | TF |
| Hydraulic cylinder | GY | TF |

# References

[1] H. Paynter, *An Epistemic Prehistory of Bond Graphs*. Amsterdam: North-Holland, 1992.

[2] D. Karnopp and R. C. Rosenberg, *System Dynamics: A Unified Approach*. New York: John Wiley & Sons, 1975.

[3] J. U. Thoma, *Introduction to Bond Graphs and Their Applications*. Oxford: Pergamon Press, 1975.

[4] P. E. Wellstead, *Introduction to Physical System Modelling*. London: Academic Press, 1979.

[5] P. C. Breedveld, "Thermodynamic bond graphs and the problem of thermal Inertance," *J. of the Franklin Institute*, vol. 314, no. 1, pp. 15–40, 1982.

[6] W. Borutzky, *Bond Graph Methodology: Development and Analysis of Multidisciplinary Dynamic System Models*. London: Springer-Verlag, 2010.

[7] C. Ma and Y. Hori, "Application of bond graph models to the representation of buildings and their use," in *Proc. of American Control Conference*, (Boston, Massachusetts, USA), pp. 2901–2906, 2004.

[8] J. J.-H. Tsai and J. S. Gero, *Unified Energy-based Qualitative Representation for Building Analysis*. Saarbrucken, Germany: VDM Verlag, 2009.

[9] J. J.-H. Tsai and J. S. Gero, "A qualitative energy-based unified representation for buildings," *Automation in Construction*, vol. 19, no. 1, pp. 20–42, 2010.

[10] J. W. Brewer, "Bond graphs of microeconomic systems," in *75-WA/Aut-8*, (Houston, USA), American Society of Mechanical Engineering, 1975.

[11] J. W. Brewer, "Structure and cause and effect relations in social systems simulations," *IEEE Transactions on Systems Man and Cybernetics*, vol. 7, no. 6, pp. 468–474, 1977.

[12] J. W. Brewer and P. P. Craig, "Bilinear, dynamic single-ports and bond graphs of economic systems," *J. of the Franklin Institute*, vol. 313, no. 4, pp. 185–196, 1982.

[13] J. W. Brewer, "Progress in the bond graph representations of economics and population dynamics," *J. of the Franklin Institute*, vol. 328, no. 5/6, pp. 675–696, 1991.

[14] Y. K. Wong, "Application of bond graph models to economics," *International Journal of Modelling and Simulation*, vol. 21, no. 3, pp. 181–190, 2001.

[15] J. A. T. Machado and M. E. Mata, "A fractional perspective to the bond graph modelling of world economies," *Nonlinear Dynamics*, vol. 80, no. 4, pp. 1839–1852, 2015.

[16] N. Guijarro and G. Dauphin-Tanguy, "Approximation methods to embed the non-integer order models in bond graphs," *Signal Processing*, vol. 83, no. 11, pp. 2335–2344, 2003.

[17] T. J. Connolly and J. A. Contréras, "Bond graph primitives for modeling systems with fractional differential equations," *Fractional Calculus and Applied Analysis*, vol. 12, no. 4, pp. 391–408, 2009.

[18] J. Deskur, "Models of magnetic circuits and their equivalent electrical diagrams," *The International Journal for Computation and Mathematics in Electrical and Electronic Engineering*, vol. 18, no. 4, pp. 600–610, 1999.

[19] A. Mukherjee, R. Karmakar, and A. K. Samantaray, *Bond Graph in Modeling, Simulation and Fault Identification*. New Delhi: CRC Press, 2006.

[20] J. Thoma and B. O. Bouamama, *Modelling and Simulation in Thermal and Chemical Engineering: A Bond Graph Approach*. Berlin, Heidelberg: Springer-Verlag, 2010.

[21] W. Borutzky, ed., *Bond Graphs for Modelling, Control and Fault Diagnosis of Engineering Systems*. Switzerland: Springer, 2 ed., 2017.

[22] J. T. Machado, "Bond graph and memristor approach to DNA analysis," *Nonlinear Dynamics*, vol. 88, no. 2, pp. 1051–1057, 2017.

[23] P. J. Gawthrop and L. S. Smith, *Metamodelling: Bond Graphs and Dynamic Systems*. Englewood Cliffs, N.J.: Prentice Hall, 1996.

[24] P. J. Gawthrop and G. P. Bevan, "Bond-graph modeling: A tutorial introduction for control engineers," *IEEE Control Systems Magazine*, vol. 27, no. 2, pp. 24–45, 2007.

[25] L. O. Chua, "Memristor - the missing circuit element," *IEEE Transactions on Circuit Theory*, vol. 18, no. 2, pp. 507–519, 1971.

[26] L. O. Chua and S. M. Kang, "Device modeling via basic nonlinear circuits elements," *IEEE Transactions on Circuits and Systems*, vol. 27, no. 11, pp. 1014–1044, 1980.

[27] M. D. Ventra, Y. V. Pershin, and L. O. Chua, "Circuits elements with memory: Memristors, memcapacitors and meminductors," *Proceedings of the IEEE*, vol. 97, no. 10, pp. 1717–1724, 2009.

[28] D. B. Strukov, G. S. Snider, D. R. Stewart, and R. S. Williams, "The missing memristor found," *Nature*, vol. 97, pp. 80–83, 2008.

[29] D. Jeltsema and A. Dòria-Cerezo, "Port-Hamiltonian formulation of systems with memory," *Proceedings of the IEEE*, vol. 100, no. 6, pp. 1928–1937, 2012.

[30] J. A. T. Machado, "Fractional generalization of memristor and higher order elements," *Communications in Nonlinear Science and Numerical Simulation*, vol. 18, no. 12, pp. 264–275, 2013.

[31] J. T. Machado and A. M. Lopes, "Multidimensional scaling locus of memristor and fractional order elements," *Journal of Advanced Research*, vol. 25, pp. 147–157, Sept. 2020.

[32] "20-sim." https://www.20sim.com/.

[33] "CAMPG (computer aided modeling program with graphical input)." http://bondgraph.com/.

[34] "Matlab®." https://www.mathworks.com/matlabcentral/fileexchange/60376-bond_graph.

[35] "The Mathematica bond graph toolbox." http://library.wolfram.com/infocenter/Conferences/4903/.

[36] *New developments in bond graph modeling software tools: the computer aided modeling program CAMP-G and MATLAB®*, 1997.

[37] J. A. Calvo, C. Álvarez Caldas, and J. L. S. Román, "Analysis of dynamic systems using bond graph method through SIMULINK," in *Engineering Education and Research Using MATLAB®* (A. Assi, ed.), ch. 11, pp. 265–288, IntechOpen, 2011.

[38] P. Šarga, D. Hroncová, M. Čurilla, and A. Gmiterko, "Simulation of electrical system using bond graphs and MATLAB®/Simulink," *Procedia Engineering*, vol. 48, pp. 656–664, 2012.

[39] C. Kleijn, M. A. Groothuis, and H. G. Differ, *20-sim 4.8 Reference Manual*. Controllab Products B.V., Enschede, Netherlands, 2019.

# Index

1-port
    dissipators, 33
    energy sources, 31
    energy stores, 32
2-ports
    gyrator, 34
    transformer, 34
20-sim, 171
    bond graph construction, 174
    frequency domain analysis, 192
    time domain analysis, 180

Belt and pulley, 15
Bond, 28
    information, 28
Bond graph, 27
    causality, 44
    elements, 30
    junctions, 35
    modeling, 27
    modulated elements, 36
    power variables, 29
    tetrahedron of state, 28

Causality
    conflict, 51, 53, 55
    dissipators, 48
    energy sources, 46
    energy stores, 47
    gyrator, 49
    indeterminate, 51, 52
    junctions, 49
    preferred, 47
    procedure to assign, 50
    transformer, 49

DC motor
    model, 19

Electrical systems
    basic elements, 3
    bond graph elements, 31–33, 58
    bond graph modeling, 57
    causality, 58
    classic modeling, 2
    proposed exercises, 66
    solved problems, 58

Faraday's laws, 2, 6

Gear train, 15
Gyrator, 6

Hydraulic cylinder, 24
Hydraulic press, 24
Hydraulic systems
    basic elements, 21
    bond graph elements, 31–33, 105
    bond graph modeling, 105
    causality, 105
    classic modeling, 20
    proposed exercises, 114
    solved problems, 105
Hydraulic transformer, 24

Junctions, 35
    0 junction, 36

1 junction, 36
  causality, 49

Kirchhoff's laws, 2, 36

Laplace transform, 1
Lead screw, 16
Lever, 18

Mechanical systems
  basic elements, 9, 12
  bond graph elements, 31–33, 76
  bond graph modeling, 75
    option 1, 75
    option 2, 76
  causality, 76
  classic modeling, 7
  proposed exercises, 86
  rotational, 11
  solved problems, 78
  translational, 8
  transmission systems, 14
Memristor, 29
  causality, 48
Multi-domain systems
  proposed exercises, 150
  solved problems, 143

Newton's laws, 8, 9, 12

Rack and pinion, 18

Thermal systems
  basic elements, 25
  bond graph elements, 31–33, 127
  bond graph modeling, 127
  causality, 127
  classic modeling, 25
  proposed exercises, 134

  solved problems, 127
Transformer
  electrical, 6
  hydraulic, 23
  mechanical, 15